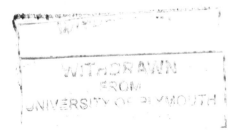

SPECIES

SPECIES AND SYSTEMATICS

The Species and Systematics series will investigate fundamental and practical aspects of systematics and taxonomy in a series of comprehensive volumes aimed at students and researchers in systematic biology and in the history and philosophy of biology. The book series will examine the role of descriptive taxonomy, its fusion with cyber-infrastructure, its future within biodiversity studies, and its importance as an empirical science. The philosophical consequences of classification, as well as its history, will be among the themes explored by this series, including systematic methods, empirical studies of taxonomic groups, the history of homology, and its significance in molecular systematics.

Editor in Chief: Malte C. Ebach (International Institute for Species Exploration, Arizona State University, USA)

Editorial Board

Marcelo R. de Carvalho (Universidade de São Paulo, Brazil)

Anthony C. Gill (Arizona State University, USA)

Andrew L. Hamilton (Arizona State University, USA)

Brent D. Mishler (University of California, Berkeley, USA)

Juan J. Morrone (Universidad Nacional Autónoma de México, Mexico)

Lynne R. Parenti (Smithsonian Institution, USA)

Quentin D. Wheeler (Arizona State University, USA)

John S. Wilkins (University of Sydney, Australia)

Kipling Will (University of California, Berkeley, USA)

David M. Williams (Natural History Museum, London, UK)

University of California Press Editor: Charles R. Crumly

http://www.ucpress.edu/scrics/spsy.php

SPECIES

A HISTORY OF THE IDEA

John S. Wilkins

UNIVERSITY OF CALIFORNIA PRESS

BERKELEY   LOS ANGELES   LONDON

University of California Press, one of the most
distinguished university presses in the United States,
enriches lives around the world by advancing
scholarship in the humanities, social sciences, and
natural sciences. Its activities are supported by the UC
Press Foundation and by philanthropic contributions
from individuals and institutions. For more information, visit www.ucpress.edu.

Species and Systematics, Vol. 1
For online version, see www.ucpress.edu.

University of California Press
Berkeley and Los Angeles, California

University of California Press, Ltd.
London, England

© 2009 by the Regents of the University of California

Library of Congress Cataloging-in-Publication Data

Wilkins, John S., 1955–
    Species : a history of the idea / John S. Wilkins.
        p.   cm. — (Species and systematics ; v. 1)
    Includes bibliographical references and index.
    ISBN 978-0-520-26085-6 (cloth : alk. paper)
    1. Species—History.   2. Species—Philosophy.   I. Title.

QH83.W527 2009
578.01'2—dc22   2009009184

Manufactured in the United States of America

16   15   14   13   12   11   10
10   9   8   7   6   5   4   3   2

The paper used in this publication meets the minimum
requirements of ANSI/NISO Z39.48-1992 (R 1997)
(Permanence of Paper).

Cover image: Detail of Ascidiae (sea squirts and
tunicates) from Tafel 85 of Ernst Haeckel, *Art Forms in
Nature: The Prints of Ernst Haeckel*. Courtesy of Prestel:
Munich, Berlin, London, New York, 1998, 2009.

Contents

Preface

The history of research into the philosophy of language is full of *men* (who are rational and mortal animals), *bachelors* (who are unmarried adult males), and *tigers* (though it is not clear whether we should define them as feline animals or big cats with a yellow coat and black stripes).

Umberto Eco [1999: 9]

"What sort of insects do you rejoice in, where you come from?" the Gnat inquired.

"I don't *rejoice* in insects at all," Alice explained, "because I'm rather afraid of them — at least the large kinds. But I can tell you the names of some of them."

"Of course they answer to their names?" the Gnat remarked carelessly.

"I never knew them do it."

"What's the use of their having names," the Gnat said, "if they won't answer to them?"

"No use to *them*," said Alice; "but it's useful to the people that name them, I suppose. If not, why do things have names at all?"

"I can't say," the Gnat replied. "Further on, in the wood down there, they've got no names."

Lewis Carroll [1962: 225]

Why look at one concept in science, out of context of the larger theories, practices, and societies in which it occurs? Why trace "species"? This sort of question is raised by both philosophers and historians when histories of scientific ideas are written.

Philosophers tend to dislike history for several reasons. One is that they often address issues and ideas as if the opponent is sitting across the symposium table from them, no matter whether that opponent lived last week, last century, or last millennium. Philosophers of science often treat history as a source of anecdotes to illustrate some more general point, such as the way the Copernican revolution changed philosophical understanding or how genes overcame vitalism. Famously or infamously,

Imre Lakatos "rationally reconstructed" the history of scientific ideas in a footnote, because history is messy and failed to clearly illustrate the philosophical point.

Historians tend to dislike intellectual histories, because such histories treat ideas as free-floating objects ("free-floating rationales," as Dennett calls them) independent of the individual psychologies and life histories, and of the social conditions in which they were raised and elaborated. Also, histories of ideas are too easy to do. All you need do is find some apparent resemblance between ideas at time *a* and time *b*, and you have a narrative. Historians, rightly, want to see actual historical influences and the effects of social and cultural contexts, the differing *epistemes* at work.

Both professions can go too far. I think history comes in a number of scales, which following a practice in ecology, I will call *alpha history*, *beta history*, and *gamma history*. Alpha history is done by investigating archives and looking at locales and artifacts. It is hard and local work, and will give the *data* of the larger-scale histories. Beta history is done by covering a restricted period, or biography, or event. It relies on the alpha material and synthesizes it into a narrative explanation of the subject. Gamma history, though, is out of fashion. Rather than being a "life and times" or "history of the period," it attempts to take alpha and beta historical work and synthesize a grand-scale narrative. And because a *really* grand-scale narrative is almost impossible to do by one person, it pays to limit the subject to something manageable. This book is at the edge (some might uncharitably say, over the edge) of that limit. But if gamma history is not worth doing, why is alpha and beta history?

Philosophy of science has become increasingly grounded in history. It is becoming the norm for philosophers of science to appeal closely to the historical development, failures as well as successes, or a given discipline or problem. Majorie Grene and Ian Hacking are perhaps the exemplars of this approach, although David Hull has also made a plea for actual examples in philosophy of biology [Hull 1989a]. And historians of science such as Polly Winsor and Jan Sapp have offered excellent case studies and narratives of all three kinds for philosophers to use. There is a shift toward this now, and that might justify a conceptual history at this time. However, there's another reason for writing this now, and that is that if philosophers don't do this, and historians don't, the scientists will, and have. A major target of this book is the scientist-developed *essentialism story* of the past fifty years. Polly Winsor and Ron Amundson, among others, have written critiques of the view that before Darwin, every biologist was held in thrall to Aristotle's essentialist biology, but there is no overall summary of this. Also, the essentialism story is used to justify or critique various species conceptions by the biologists themselves. History has a role in scientific debate.

Generally, scientists have a "rolling wall of fog" that trails behind them at various distances for different disciplines, above which only the peaks of mountains of the Greats can be seen. In medical biology, for instance, this wall is about five years behind the present. Little is cited before that, and those works that are, are cited by nearly everyone. So there is a tendency for what Kuhn called "textbook history" to become the common property of all members of the discipline. However, taxonomy is an unusual discipline, in that the classical works are more widely cited and appealed to than in most other sciences. The ideas of an eighteenth-century Swede or French author can carry weight in a way that the genetics or physics of that time do not. Partly this is because a large element of taxonomy is conceptual: logical and metaphysical ideas, which change slowly, carry probative force. So asking "What is a species?" is to ask a historical as well as a present question, and how the notion of species arrived at the present debate in part defines that debate.

Doing this kind of history is rather like trying to work out the past from a series of old photographs in a box in the attic you got from your grandparents. Faces appear in various guises, resemblances recur, and it is almost impossible to identify exactly who is whose child, friend, or mere passers-by. Nevertheless, having that box of snapshots, one is richer for it in understanding both the past and the present.

So I seek absolution from each of these three professions—philosophy, history, and biology. I believe I can show there is a basic error involved in the essentialism story that can be resolved by a conceptual history. Scientific history is at least partially conceptual, so I don't think it is illicit to write a conceptual history. But the conceptual history of an *idea*? That might be too much. Well, this is not exactly the history of an idea. It is a combined history of *various* ideas and words that have a subtle ambiguity in philosophy *and* biology. And it is my claim that this ambiguity has confused the present debate over species in both fields. I will summarize the argument here, so that it is clear what the issues this book addresses are.

The essentialism story is a view that has taken biologists and philosophers by storm. Primarily advocated, and largely developed (I hesitate to say "invented") by Ernst Mayr, it is the view that there are basically two views of biological taxa in general and species in particular. One is the view deriving from Plato and Aristotle, on which all members of a type were defined by their possession of a set of necessary and sufficient properties or traits, which were fixed, and between which there was no transformation. This is variously called *essentialism, typological* or *morphological thinking*, and *fixism*. The other is a view developed in full by Charles Darwin, in which taxa are populations of organisms with variable traits, which are polytypic (have many different types) and which can transform over time from one to another taxon, as the species that comprise them, or the populations that comprise a species, evolve. There are no necessary and sufficient traits. This is called *population thinking*.

I will argue that the essentialism story is false. It is based, as Polly Winsor has shown, on a misreading of the *logical* tradition in which *species* is a class that is differentiated out of a larger class, or *genus*, by a set of necessary properties all members bear, as being a claim about *biological* species, which are not so defined but are instead diagnosed by morphological characters. There is a clear distinction between the *formal* definitions of logical species and the *material* characters and powers of the biological organisms of a biological species pretty well from the beginning of modern natural history, but arguably even from Aristotle onward. This is the first claim.

A second related claim is that living species were always understood to include or require a *generative power* rather than morphological similarity or identity, which was always held to be a way of identifying them at best. I call this the *generative conception of species*, and it was held as much in the classical writings of Aristotle and Theophrastus, as by the moderns, and the present views of species are equally within that ancient tradition. There are plenty of cases of medieval, early modern, and recent pre-Darwinian authors employing some variant of this conception, allowing for deviation from types. Moreover, the use of diagnostic characters neither requires essences of a causal or material kind, nor implies that all diagnostic characters are borne by every member of the species. In the diagnostic sense, Darwin was as much a formal essentialist as Linnaeus, and modern taxonomists are still today. In short, generation, not definition, is what counts for living species.

My third claim is that the notions of fixity of species and essence and type are decoupled from each other. Types are not the same as essences, and they had a number of roles to play in classification. Fixity of species was invented by John Ray in the seventeenth century, and repeated by Linnaeus, and it was solely based on piety, not metaphysics. Where material essences were employed—for instance, by Nehemiah Grew—they were considered to be causal essences for biological structures rather than of taxa, and this was also the view of the ideal morphologists of the early nineteenth century after Goethe. If essences play any role in taxonomy at all, it must be well after Darwin, and it is possible it never was really held before the essentialist story developed, or was invented, around 1958.

So much for the history. Philosophically, essentialism underwent a revival in the 1960s in the philosophy of language, in part because of a reaction to Popper's philosophical attack on *methodological* essentialism outside biology. This was, I believe, a reason why essentialism became a problem in philosophy of biology. Platonism, however, is a recurring theme in both biology and philosophy in the scientific period, and it is possible that philosophical interpretations of biology in the period were motivated in part by the increasing atomism and materialism of writers like Locke and Boyle, who were influential on many biologists, such as Buffon and Lamarck.

There is a distinction to be had between the naming of species and their underlying biological descriptions. Part of the modern debate rests on a confusion between nomenclature and the reference of names, on the one hand, and the accounts given of the formation and maintenance of species, on the other. Species "nominalists," whom I refer to as species deniers in the biological traditions, deny that the names refer to anything biologically real (which is not to argue against scientific realism—the reality referred to here is simply the status of species as objects in biology; whether they are real in a metaphysical sense is outside our scope here). Species taxa realists hold that the taxa are real objects in biology. Species *category* realists also believe that the *rank* of species is a real phenomenon, in that species have a unique and peculiar organization that other taxa, below and above that rank, do not have.

In summary, then, we have three claims that this book is intended to demonstrate: the logical and the natural species are distinct ideas that largely share only a term; there was a single species "concept" from antiquity to the arrival of genetics, the *generative conception*; and types are neither the same as essences nor something that changes much with Darwin.

The book also lists the modern "species concepts" (or, rather, definitions of the word and concept *species*) in play. It will become clear that most of these are partial definitions, some of which are simply not operationally applicable. It is my belief that all biologists doing systematics will construct for themselves a conceptual sandwich out of these elements, one that suits the tastes required for studying the group of organisms they do. So the latter part of the book is a kind of "conceptual delicatessen." Biologists, like anyone else engaged in an intellectual enterprise, tend toward monism of theoretical ideas: so *species* conceptions that work for them must work in all other cases. This is especially obvious in the "animal bias" of workers like G. G. Simpson and Theodosius Dobzhansky, who simply denied that things that were not species the way sexual animals were species, were in fact not species at all. I hope to show here that the term has a wider application than in a few leading cases. The philosophical arguments, however, must wait for another book.

Acknowledgments

This book has taken me the better part of a decade to write, and I have done so with the aid of many people. I am consistently surprised and pleased to find that even those individuals with whom I have great disagreements over species concepts turn out to be extremely nice, helpful, and above all scholarly individuals. I never met Mayr, but those I know who have say the same thing about him. So, I am deeply indebted to the following for criticism, editing, material, suggestions, and the occasional expression of outright incredulity:

Gareth J. Nelson and Neil Thomason were all that a doctoral candidate could desire of thesis advisors, and their critical comments have helped make my thesis, from which this book evolved, a much better work.

Thanks to Quentin Wheeler, Malte Ebach and Chuck Crumly for editorial advice and help.

Thanks also to, in alphabetical order, Noelie Alito, Mike Dunford, Malte Ebach, Greg Edgcombe, Dan Faith, Michael Ghiselin, Paul Griffiths, Colin Groves, John Harshman, Jody Hey, Jon Hodge, David Hull, Jon Kaplan, Mike S.-Y. Lee, Murray Littlejohn, Brent Mishler, Larry Moran, Staffan Müller-Wille, Ian Musgrave, Gary Nelson, Mike Norén, Gordon McOuat, Massimo Pigliucci, Tom Scharle, Kim Sterelny, Neil Thomason, Charissa Varma, John Veron, Quentin Wheeler, David

Williams, and Polly Winsor, who all provided information, criticism, advice, and assistance, some considerable.

I also thank members of the audiences at the 2000 Australasian Association for the History, Philosophy and Social Studies of Science Association conference at the University of Sydney and the 2001 International Society for the History, Philosophy and Social Studies of Biology conference in Hamden, Connecticut, and the Systematics Forum at the Melbourne Museum, run by Robin Wilson. Michael Devitt also forced me to reassess the nature of essentialism at a later workshop, although without much success.

Polly Winsor, David Hull, Joel Cracraft, Norman Platnick, and Ward Wheeler graciously granted time for an interview, and Joel has been of particular help with the phylogenetic conception. I must especially thank the late Herb Wagner, John Veron, Mike Dunford, Tom Scharle, and Scott Chase for technical information provided. Other acknowledgments are made in the notes. I sincerely apologize to anyone I have left unjustly unthanked.

I owe a debt to the Walter and Eliza Hall Institute of Medical Research in Melbourne, which was my employer during my PhD candidacy, and to the Biohumanities Project (Federation Fellowship FF0457917), under the direction of Paul Griffiths, at the University of Queensland, where the bulk of the work on the Late Middle Ages and Renaissance, and some of the work on the eighteenth century was done, with the assistance of the library there. I must make special mention of Gallica project at the Bibliothèque nationale de France (gallica.bnf.fr), the Internet Archive (www.archive.org), Botanicus (www.botanicus.org), and the Complete Works of Charles Darwin project (darwin-online.org.uk), overseen by John van Whye. Without these resources many texts would have remained inaccessible to me in Australia.

Therefore, in the light of this gracious assistance from so many people, it follows that all misunderstandings, errors, and incoherencies that remain—and there will be many—are my own fault.

PROLOGUE

Progress, far from consisting in change, depends on
retentiveness. Those who cannot remember the past are
condemned to repeat it.

George Santayana [1917: 284]

First, then, we have the problem involved in the origin of
species. As a preliminary to that, logic demands that we
should define the term. It may be that logic is wrong, and
that it would be better to leave it undefined, accepting the
fact that all biologists have a pragmatic idea at the back
of their heads. It may even be that the word is undefinable.
However, an attempt at a definition will be of service in
throwing light on the difficulties of the biological as well as
of the logical problems involved.

Julian Huxley [1942: 154]

THE RECEIVED VIEW

The problem of species is a long-standing one in biology. Since the
development of modern taxonomy and classification, naturalists,
botanists, zoologists, and all the other various terms for aspects of biology as we have known it or now know it have tried to define clearly what
it is they are taking about when they talk of "species." There is a Received View of the species concept that is largely the result of work published by biologists themselves.[1] Polly Winsor [2003] traces the origin of
the Received View history to Arthur Cain's [1958] reliance on a 1916
"misinterpretation" of Aristotle [Joseph 1916; cf. Winsor 2001][2] and

subsequent writings on the history of taxonomy [Cain 1959a, 1959c]. These influenced both Simpson and Hull, and also Mayr's later development of the story. Similar conclusions have also been reached by Ron Amundson concerning typology [Amundson 1998, 2005]—transcendentalism and idealism were less essentialistic than the Received View indicated, a view in part revised by Hull himself [Hull 1983b]. The Received View appears in fragmentary form throughout a number of publications over the past sixty years, and we shall look at three examples.

George Gaylord Simpson was perhaps the greatest of the paleontological evolutionary theorists of the twentieth century. His influential text on animal taxonomy [Simpson 1961] included a chapter entitled "The Development of Modern Taxonomy." According to Simpson, Scholastic logic, based on Aristotle, relied on the "essence" of a "species," which consists of its "genus" plus its "differentia," which was a logical relationship [p. 36f.]. It was this that Linnaeus adopted [p. 38], and it has several serious faults: one is that "properties" and "accidents" (Simpson's scare quotation marks) are excluded from the "essence" *and therefore the definition* (my emphasis) of the species. Another is that this method produces a classification of *characters*, not *organisms* (his emphasis). From Linnaeus to Darwin, there was an increasing emphasis on empirical rather than a priori classification.

According to Simpson, there was a typological tradition stemming from Plato and coming via Aristotle, neo-Platonic, Scholastic, and Thomist philosophy into biological taxonomy [p. 47]. He writes:

> The basic concept of typology is this: every natural group of organisms, hence every natural taxon in classification, has an invariant, generalized or idealized pattern shared by all members of the group. The pattern of the lower taxon is superimposed upon that of the higher taxon to which it belongs, without essentially modifying the higher pattern. . . . Numerous different terms have been given to these idealized patterns, often simply "type" but also "archetype," "Bauplan" or "structural plan," "Morphotypus" or "morphotype," "plan" and others.

In contrast, modern taxonomy relies on common descent [pp. 52–54], statistical properties of populations [p. 65], and biological relationships. Simpson took his historical information on this subject from a paper by A. J. Cain [1958; cf. Winsor 2001]. Cain, following the interpretation of Aristotle and the later logicians of H. W. B. Joseph [1916], had said that Linnaeus based his conception of species on Aristotelian definition of essences, on "*a priori* principles agreeable to the rules of logic" [p. 147, quoted in Winsor 2001: 249]. He later reversed this position after retirement and time to properly read Linnaeus and Ray [e.g., in Cain 1993, 1994, 1999a, 1999b]. Nevertheless, his earlier papers [Cain 1958, 1959a, 1959b] became the foundation for the Received View.

Ernst Mayr is someone whose ideas and narratives have been extremely influential, and therefore with whom we will deal later in some detail. He has often put forward a narrative history of species concepts, and his ideas are the most widely known and accepted. According to him, the species concept begins in biology with Linnaeus [Mayr 1957: 2f.] since before then, apart from Ray, nobody believed species were stable entities. Later Darwin held that species were fluid [p. 4], but in the intervening period, there was a tendency of various writers like Cuvier, De Candolle, Godron,[3] and von Baer to treat them as real and definite entities, united, as von Baer says, by common descent from original stock. The realization that species exhibited a "supraindividualistic bond" [p. 8] and that members of species reproduce only with each other came slowly. The older view was "typological," and it is "the simplest and most widely held species concept" [p. 11]. Typology, according to Mayr, as for Simpson, is due to the influence of Plato, and those who follow him are trying to define a species in terms of "typical" or "essential" attributes [p. 12]:

> Typological thinking finds it easy to reconcile the observed variability of the individuals of a species with the dogma of the constancy of species because the variability does not affect the essence of the eidos [the Greek term translated as "species"] which is absolute and constant. Since the eidos is an abstraction derived from human sense impressions, and a product of the human mind, according to this school, its members feel justified in regarding a species "a figment of the imagination," an idea.

In contrast, he says, a species today is regarded as a gene pool rather than a class of objects.

Mayr presents this history in more detail in his *The Growth of Biological Thought* [Mayr 1982: 254–279]. Here again, the distal source of essentialism is Plato, now via Aristotle, to John Ray and Linnaeus. Again, it is only with the increasing emphasis on empiricism and the development first of evolution and then of genetics that naturalists begin to realize that species are gene pools, reproductively isolated from each other. In the end, the biological species concept is triumphant.

A third history, for contrast, is that provided by David Hull, who became the leading philosopher of biology of his generation. He made it a point to focus on the actual history and biology of his subjects rather than on a "rational reconstruction" or textbook history, as was sometimes the practice of prior philosophical scholarship of science. In his seminal paper "The Effect of Essentialism on Taxonomy: Two Thousand Years of Stasis" [Hull 1965], he presented the Received View, in part relying on Joseph's book,[4] in part on Mayr and Simpson. The foundations of classification rest on Aristotle's notion of definition [p. 318f.], while a contrary tradition, beginning with Adanson, treated taxa as disjuncts of properties, which themselves became the essence of a species in classical Aristotelian form. The moderns—Simpson, Dobzhansky, Mayr—moved to a reproductive isolation conception, founded on evolution. In a much later work, he devoted a chapter [chapter 3, "Up from Aristotle," Hull 1988c] in which the Received View is expanded and in many ways revised: Plato is relegated to the background, and Aristotelians offered a view of species as types from which deviations were possible. After Aristotle, we reach Linnaeus, then Buffon, Lamarck, Cuvier, Geoffroy, and then the ideal morphologists until we meet Darwin, followed by the New Systematics group—Dobzhansky, Mayr, Julian Huxley, and so forth.

In each of these, and others [e.g., Wiley 1981: 70–72, to take one example at random], the Received View runs roughly thus:

> Plato defined Form *(eidos)* as something that had an essence, and Aristotle set up a way of dividing genera *(genē)* into species *(eidē)* so that each species shared the essence of the genus, and each individual in the species shared the essence of the species. Linnaeus took this idea and made species into constant and essentialistic types. Darwin overcame this essentialism. Later naturalists, under the influence of genetics, discovered the biological species concept, in which species are found to be populations without essences, but with common ancestry. Population thinking replaces typological essentialism.

THE FAILURE OF THE RECEIVED VIEW

In between Aristotle (d. 322 BCE) and Linnaeus's first works on classification (fl. 1750 CE) there lay a two-thousand-year gap. What happened in that intervening period? It was a period of active philosophy, as it includes the bulk of the classical period, not to mention the high period of Arab science and scholarship,[5] the medieval debates on logic and universals, and the Renaissance. Are we to think that there was a change or development in these ideas in that time? Accepting this, I undertook to investigate this gap, expecting no more than a continuous narrative supporting the Received View. The results were surprising. In most respects, the Received View is incorrect, seriously incomplete, or simply false. Effectively every philosophical issue raised about biological species in the modern period was raised in one form or another about philosophical, or logical, species during this interregnum. Moreover, the crucial mediate link between Aristotle and modern biology was not Linnaeus or even Aristotle's own writings. It was, rather, the late classical neo-Platonists, rediscovered by the Cambridge Platonists (themselves properly considered also to be neo-Platonists after the school of Plotinus and Porphyry, to whose works they had direct access), by way of their influence on John Ray, John Locke, and various other seventeenth-century notables. In effect, the species problem is a neo-Platonist plot, not an Aristotelian one.

Furthermore, the Received View tended to ignore the Great Chain of Being and the universals debates, which are of great importance in the way later writers such as Cusa deal with the notion of essence and definition. It is *not* the case that typology and essentialism were bound together, and in many ways typology as it was actually discussed is more in line with Mayr's "population thinking" than he ever admitted. Winsor [2003] refers to the typological conception as the "method of exemplars," which we shall see is a much better characterization.[6]

Few truly novel *formal, conceptual* elements of species concepts have arisen since the eighteenth century. Biologists and philosophers have been dealing from the same deck of conceptual cards ever since. Of course, many *technical* concepts are novel—for example, Templeton's genetic concept [Templeton 1989] relies on Mendelian genetics, and Wu's genic concept [Wu 2001] relies on molecular genetics, both of which are much later additions to the repertoire of biology.[7] It is also open to doubt that ideas that are merely formally similar are in fact "the same" ideas as later ones. But the similarities are themselves instructive; if an older debate dealing with formally similar notions is resolved into a few opposing viewpoints, we may learn from them what to expect in modern debates.

Despite popular expectation, Darwin and his successors did not add much to the species debate except to raise in sharp relief problems brought about by the introduction of the notion of speciation and the subsequent mutability of classifications. Nevertheless, Darwin acts as a focal point for what follows, and even the formalists had to address his conceptions. Since the "modern synthesis" of 1930–1942, the only truly novel *philosophical* ideas about species have been the Individuality Thesis and the refinement of notions of class inclusion and hierarchies, which themselves rest on prior work (although I keep finding people expressing the view that species are individuals in the logical tradition—for example, Boëthius, Aquinas, and Cusa—and in the biological tradition, even

Mivart discusses a version of it, in which species are like organisms).
So far as I can tell, there are precursors or forerunners for everything
else, at least in the sense of resemblance, if not direct and demonstrated
descent. It is often the case that these ideas are continually reinvented,
especially by biologists. Equally, however, it is often the case that these
modern alternatives are not directly descended from the precursors,
although there is evidence of indirect descent. For example, Rutgers
geneticist Jody Hey recently expressed the view [Hey 2001a, 2001b] that
species are merely conventional objects used for communication, a rein-
vention in some respects of John Locke's earlier conventionalism (as we
will discuss). He depends extensively upon W. V. O. Quine's philosophy
of language in *Word and Object* [Quine 1960]. Quine is, of course, dealing with the same issues of the empiricist, linguistically directed, philosophical tradition that Locke began, and he knows his Locke very well,
but Hey understandably seems unaware of it. As a geneticist he can be
excused for not knowing the detailed history of an abstruse episode in
philosophy, but this pattern is repeated even when the originator of a
view—for instance Hennig—*is* aware of his philosophical precursors
(including in this case Woodger and Gregg[8]). Later followers of that initial account often do not realize that there ever was such a prior history.

For this reason, it is important to at least sketch some of the main pre-
biology and early biology historical accounts of species, and this of course
means beginning in the Western tradition with Plato and Aristotle. I seek
to be excused for this Western bias, despite much interesting literature
in Persian and Asian cultures on classification of organisms, because it
is from the Western tradition and not the others that the modern species
problem derives and on which it depends. For the same reason, I will
pass over the work of cultural anthropologists on classification in non-
Western indigenous cultures [e.g., Atran 1985, 1990, 1995, 1998, 1999;
Bulmer 1967; Ellen 1993]. Likewise, I shall not attend to the work done
on the developmental and psychological origins of taxonomic concepts
[Eco 1999; Estes 1994; Gil-White 2001; Griffiths 1997; Hey 2001a; Keil
1995; Millikan 1984; Sperber, Premack, and Premack 1995; Sperber and
Wilson 1986].[9] I assume here that scientific concepts, whatever their
etiology in our biological substrate, are primarily subject to change due
to the institutions and culture of science and cultural influences such as
philosophy. We all share whatever psychological and cognitive foundations there are of classification and categorical notions; but the scientific
debate has gone its own way, and so we may take it that biological biases

are not determinants of such categories. It is my view that cultural evolution, including the evolution of science [Hull 1988a, 1988c], has its own dynamics, perhaps biased by psychology and evolutionary adaptations and heritage, but that it is usually decoupled from them [Campbell 1965; Toulmin 1972].

This is an essay in the history of ideas (although I prefer the term *conceptual history*), and in particular of the ideas that came before and might be demonstrated or fairly thought to have contributed to the ideas in play in biological thought about species and classification. "History of ideas" has become uncommon and somewhat disparaged by professional historians, and this is understandable given the whiggish, presentist bias much of it has exhibited. I am well aware of the problems faced by the historian of ideas, as described by John Greene [1963: 11]: "Of all histories the history of ideas is the most difficult and elusive. Unlike things, ideas cannot be handled, weighed, and measured. They exert a powerful force in human history, but a force difficult to estimate."

But history of ideas is necessary if we are to understand why ideas are as they are in internal terms—that is, in terms of the *content* of the ideas. There is also a need for an externalist history of ideas, of the *context*, but this is much harder to do and in any event cannot be done without at least some awareness of the internal history. Without defending this further, the history of ideas is at least intrinsically interesting for those seeking to explain current scientific ideas, given that science, as an intellectual set of traditions, relies on the notions of the past as a starting point in the way it develops them further. Knowing the past may also help scientists to avoid repeating it unnecessarily.[10]

We shall not refer much to the usual labels and banners applied to the various thinkers—terms such as *idealist, transcendentalist, empiricist,* and so on. This is because it is my experience that such absolute distinctions ride roughshod over the complexities and similarities of these thinkers. Calling somebody a "transcendentalist" implies they were interested only in Platonic ideas. In the case of a Cuvier or an Agassiz, that is a hard claim to sustain—they attended closely to empirical evidence, and for all their shortcomings, neither merely made inferences *sub specie aeternitatis*, as Plato did. And in abandoning these hard distinctions, I have found, I believe, a similarity of conception I call *the generative conception of species* that runs through most, if not all, pre-Darwinian thinkers and researchers on the topic of *species*, and which is found even in post-Darwinians and Darwin, as well. As Amundson notes [1998: 159], we have a "conceptual tangle" when we attend to the actual history of concepts, although he doesn't keep his own counsel in his later book [Amundson 2005].

As always in the history of ideas, much of the earlier material must be drawn in terms of resemblance and succession. Because one author—say, Porphyry—discusses ideas similar to another's—say, Aristotle—is not in itself reason to believe, either, that there is a direct relationship of intellectual ancestry or even an indirect relationship unless the one cites or refers to the other (as Porphyry does Aristotle). Classical authors often failed to provide a proper modern scholarly apparatus of citations and imputed credit. Prior to the "biological problem" of species beginning with Bauhin, Cesalpino, and John Ray, this lineal descendency must therefore be taken as an approximation. With that caveat, let us proceed.

THE CLASSICAL ERA: SCIENCE BY DIVISION

The history of the species concept can be divided into a prebiological and a postbiological history, which is how the Received View has always treated it. But the two histories overlap substantially, and it is much better to consider instead the history of the species idea that applies to any objects of classification—the tradition of universal taxonomy and philosophical logic—and, independently, the particular history of the species idea that applies solely to biological organisms. Even though, for example, Linnaeus [Linne 1788–1793, vol. 3] famously applied the notion *species* to minerals as well as organisms, his biological usage included elements not included in the mineralogical case. We must separate universal and biological taxonomic notions of species.[1]

So we will distinguish between two kinds of taxonomy. The *universal taxonomy* is largely the philosophical tradition from Plato to Locke (but which continues through to the considerations of sensory impressions, or *qualia*, by the logical positivist and phenomenological philosophical schools) and in which species are any distinguishable or naturally distinguished categories with an essence or definition. Then there is the *biological taxonomy* that develops from this tradition as biology itself develops from the broader field known as "natural history."[2] These biological notions of species do not necessarily refer to reproductive communities, nor do they in the medical definitions of species of diseases of the period [Cain 1999b], but we do need to recognize that "species" develops a uniquely biological flavor around the seventeenth century.[3]

These two conceptions form what might be regarded as part of or an entire research program, in Lakatos's sense. The universal taxonomy program that began with Plato culminates in the attempt to develop a classification of not only all natural objects but all possible objects. It continues in some sense in the modern projects of metaphysics and modal logics. The biological taxonomy project that developed out of it resulted in a program to understand the units of biology, to which Darwin contributed, and to which genetics later added. We might see *species* as a marker for both projects. Seen in that way, we might reasonably ask what the core assumptions of these projects are. In the universal case, it is classification (and science) by division. In the biological instance, it is, as I will argue, the marriage of reproduction or generation, with form, which I call the "generative conception." This remains even today the basis of understanding *species*.

The general philosophical tradition of considering classification to be an indispensable aspect of science seems to end with Mill and Whewell (a point noted by Hull, personal communication), and although I have found later discussions on classification in Peirce and as late as 1873 in Jevons [1878], the revivification of taxonomy in philosophy of science in recent decades appears, with one exception [Woodger 1937], to be driven by biological systematics itself rather than through philosophy motivating biology. As a result, the philosophical foundations of taxonomy are critically incomplete, and systematists often rely on philosophers who disregard the matter almost entirely, like Popper [who only mentions it dismissively in Popper 1957b §27; Popper 1959: 65; as noted by Hull 1988c: 252].

TERMS AND TRADITIONS

The modern and medieval word *species* is a Latin translation of the classical Greek word *eidos*, sometimes translated as "idea" or "form." Other significant terms are also translations of Greek terms (see table): *genus* from *genos*, *differentia* from *diaphora*. Liddell and Scott [1888] tell us that *eidos* means "form," and is derived from the root word "to see," and *genos* means "kind" and is derived from the root word "to be born of." We still find these senses in the English words *specify*, *special*, *spectacle*, *generation*, *gene*, and *genesis*. *Diairesis* or "division" is an interesting term that will play a major role in our story. Another Greek term that is later adopted by the Latins is *synagoge* or, in Latin, *relatum*. *Relata* are those features by which things group together. *Essentia*, the Latin

SOME GREEK TERMS AND
THEIR TRANSLATIONS

Greek word, plural form	Classical meaning [Liddell and Scott 1888]	Latin translation	English translation
Eidos, eidē	That which is seen, form, shape, figure; a class, kind or sort.	*Species, forma*	Species, idea, kind, sort (Locke), form
Genos, genē	Race, stock, family. A generation.	*Genus* (pl. *genera*)	Genus, kind
Diaphora	Difference, distinction, variance, disagreement	*Differentia* (pl. *differentiae*)	Difference
Diairesis	A dividing, division	*Divisio*	Division
Synagoge	A bringing together; a conclusion	*Relatum* (pl. *relata*)	Relation, affinity

term used for "essence," is a neologism for a phrase used by Aristotle, the "what-it-is-to-be." The Latin term *substantia* is occasionally translated as "essence" in natural historical works, but this is a mistake that can mislead.

However, merely because words derive their etymology from older terms or translate words in other languages, it does not immediately follow that they are the same terms with the same intension or extension, or that one author is influenced by another. By *eidos* and *species*, for example, different authors have meant *forms, kinds, sorts, species* [in the logical sense of nonpredicables; Ross 1949: 57], *biological species, classes, individuals,* and *collections,* both arbitrary and artificial, or natural and objective. In particular, *species* has meant both varieties of ideas and sense impressions, *species intelligibilis* [Spruit 1994–1995], and also the material form of the elements of the sacraments in the Roman Catholic tradition. This must be borne in mind, or it will cause confusion when we consider the views of different authors. This is an instance of a more general problem, called "incommensurability" after Kuhn's thesis that terms in scientific theories can have different referents in the shift from one theory to another; Sankey [1998] has called this particular problem "taxonomic incommensurability." We should not make too much of this, but the terms used in classification shift in subtle and major ways that sometimes obscure the views each author is presenting. We are primarily concerned with the tradition outside biology that *has* impacted on the biological notions and usage. The English plural of *species* singular is *species*. The word *specie* refers to small coinage. However, in Latin there is a singular *(species)* and plural form *(speciei)*, which is signified in one text [Porphyry 1975] by italicizing the ending thus: "spec*ies*." This is clumsy for the purposes of this book, so here I will follow the rule of the biological writers of the past century and refer to "the" species or italicize the entire word for the concept and leave the term unqualified for number except by context.

For the period from the Greeks to the beginnings of natural history in the sixteenth and seventeenth centuries, it may help to follow Locke's suggestion (discussed later), to get a feel for the meanings and avoid anachronistic interpretations, by replacing *genos* and *genus* with "kind" in English, and *eidos* and *species* with "sort." Thus, the informal usage of, say, Theophrastus in talking indifferently about *genē* and *eidē*, can be seen as the same sort of informal usage an English speaker might make by stylistically mixing "kind" and "sort" in a discussion to avoid repetition. It is imperative to remember that these were not technical terms of biology until the modern period, in particular after Linnaeus.

PLATO'S *DIAIRESIS*

As the Hellenic world began the process of expanding and consolidating, early Hellenic philosophers, known as the Milesians, had in the sixth century BCE begun to grapple with the problem of change—of generation and corruption—and what that meant for knowledge. This was essentially the core problem of Greek philosophy, for on it rested the entire conception of the possibility of the knowledge of Nature *(phusis)*.

Anaximander argued that the world was composed of a single eternal substance, the *apeiron* (the unbounded), and that the world was only superficially changing. The Pythagoreans presumed that the foundation of the world was Number, and a table of contraries described by Aristotle in the *Metaphysics* 986a15 (limited-unlimited, one-many, odd-even, light-dark, good-bad, right-left, straight-curved, male-female, square-oblong, rest-motion) accounted for the world [Brumbaugh 1981: 36]. Later, Heraclitus famously asked if you could step into the same river twice, as the river changed from step to step (although he is often misunderstood to answer in the negative; rather, he thought there was something eternal that preserved reference, noting that "Nature loves to hide"; fragment 10). And so on. Various philosophers attempted to divide the world into its constituent elements and the eternal forms of reality. This led to the idea of classification, through the uncovering of the eternal *logos*, a term or class or order.

Classification was at this time uncritically applied to all things, whether artificial or natural (a distinction the early Greeks would not have fully accepted anyway) and whether conceptual, semantic, or empirical. The problem of how to properly classify things, including living things, is first recorded to be dealt with in detail by Plato in the *Sophist* [219a–221a]. Plato, regarded by many later thinkers, particularly in the Renaissance, as *The* Philosopher, founded the philosophical school known as the Academy in Athens. He proposed a method of binary division of contraries until the object being classified was reached; this was known as the *diairesis* (division or, as some call it, dichotomy), in a similar fashion to the Pythagoreans.[4] For example, he somewhat whimsically defined fly-fishing as a model for all such classification. He has the Stranger of the dialogue ask leading questions, such as whether the fisher has skill (*techne* = art) or not, defining art into two kinds: agriculture and tending of mortal creatures, on the one hand, and art of imitation, on the other. Later he introduces a fairly arbitrary distinction about acquisitive art, resulting in the following "final" definition of angling:

> *Stranger:* Then now you and I have come to an understanding not only about the name of the angler's art, but about the definition of the thing itself. One half of all art was acquisitive—half of the acquisitive art was conquest or taking by force, half of this was hunting, and half of hunting was hunting animals, half of this was hunting water animals—of this again, the under half was fishing, half of fishing was striking; a part of striking was fishing with a barb, and one half of this again, being the kind which strikes with a hook and draws the fish from below upwards, is the art which we have been seeking, and which from the nature of the operation is denoted angling or drawing up *(aspalieutike, anaspasthai)*. [Jowett's translation]

As Oldroyd summarizes it [1986: 42], for Plato "angling is a coercive, acquisitive art, carried out in secret, in which live animals living in water are hunted during the day by blows that strike upwards from below!" (figure 1). In addition to division, Plato also classified by grouping *(synagoge)*, so that he divided things and grouped them according to their differences and similarities [Pellegrin 1986]. Plato's classification style

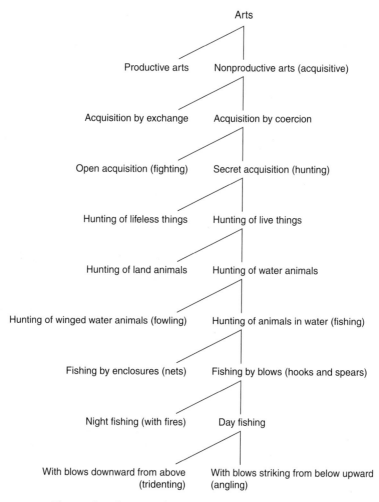

Arts
Productive arts — Nonproductive arts (acquisitive)
Acquisition by exchange — Acquisition by coercion
Open acquisition (fighting) — Secret acquisition (hunting)
Hunting of lifeless things — Hunting of live things
Hunting of land animals — Hunting of water animals
Hunting of winged water animals (fowling) — Hunting of animals in water (fishing)
Fishing by enclosures (nets) — Fishing by blows (hooks and spears)
Night fishing (with fires) — Day fishing
With blows downward from above (tridenting) — With blows striking from below upward (angling)

Figure 1 Plato's classification of angling

here is clearly arbitrary. In order to force the division into dichotomies, he (through his "Stranger") selects the "right" connections for the next differentia, but nothing is obvious about these steps, and it is clear that he knows ahead of time what he wants to deliver. In short, this is question begging, a party trick not unlike his "showing" that a slave boy "remembered" the proof of Pythagoras's theorem in the *Meno* [82b–85b].

Plato does not distinguish between classification of natural things and artificial things. For him (or, rather, his protagonist Socrates), issues of justice and social order are on a par with, and in fact transcend, issues of the natural world. Plato shows little deep interest in any aspect of the terrestrial world [Kitts 1987], although the heavens, being as close to the eternal as possible, are recommended as philosophical objects of study [*Timaeus* 27d–34b].

Clearly, there is considerable detail to Plato's views of Form or Idea *(eidos)* that must be passed over here, but this is a well-understood field of history of philosophy, and as it is entirely outside the scope of the present treatment, I am forced to avoid it.[5] For Plato the *ideai* were metaphysical or ontological realities that were neither changeable nor in the transitory world and that were preconditions for knowledge. This is the commencement point for the later tradition of forms as the basis for classification we find in Aristotle—for instance, the famous "carve nature at its joints" passage in the *Phaedrus* 265d–266a:

> *Socrates:* The second principle is that of division into species according to the natural formation, where the joint is, not breaking any part as a bad carver might. [Jowett's translation][6]

Socrates then says he is a great lover of division and generalization, and follows anyone who can see the "One and the Many" in nature. Note, however, that Socrates includes human society in the term *nature* here.

Nelson and Platnick [1981: 67f.] quote a passage from Plato's *The Statesman* [262d] in which Plato presents an argument against incomplete groups that is similar in many ways to Aristotle's argument against privative groups given in the next section, but this point is not mediated to later writers on logic. As Hull [1967: 312] observes, Plato's direct influence on biology is late, not until the seventeenth century. Indirectly, though, is another matter, as I will argue later.

ARISTOTLE: DIVISION, AND THE GENUS AND THE SPECIES

If, as Whitehead said, Western thought is a series of footnotes to Plato, then biological thought is a series of footnotes to Plato's onetime pupil Aristotle, a fact also noted by Darwin shortly before his death, when, in a letter to William Ogle thanking him for a copy of his translation of Aristotle's *Parts of Animals* on February 22, 1882, he wrote, "From quotations which I had seen I had a high notion of Aristotle's merits, but I had not the most remote notion what a wonderful man he was. Linnaeus and Cuvier have been my two gods, though in very different ways, but they were mere school-boys to Aristotle. I never realized before reading your book to what an enormous summation of labor we owe even our common knowledge."[7]

Aristotle traveled widely throughout the Hellenistic world, tutoring the young Alexander before he became Great. He therefore lived at a time of increased travel and trade, as well as during the flowering of Greek thought and science. As such, he was able to access a great deal more information than had, for instance, the Presocratics. As Hull [1988c: 75–77] notes, we may quibble over whether Aristotle was a scientist, but that he behaved like one is not at issue. A large part of Aristotle's poor reputation results from the rhetoric of the Renaissance humanists, who sought to downplay the worth of the leading contemporary source of Scholastic philosophy and theology.

Aristotle wrote several works of a biological nature, the most prominent for our purposes being *On the Parts of Animals*, *The History of Animals*, and *On the Generation of Animals*, which came in the later medieval period, after being translated into Latin, to be known as the *Liber Animalium*.

In his formal works, he employed the logical notions *genos* (genus), *eidos* (species), and *diaphora* (differentia) [Pellegrin 1986]. These were not special biological notions; in fact, they were part of his project of a wider classificatory logic, outlined in the *Metaphysics*, the *Categories*, and the *Posterior Analytics*. However, Aristotle was not an abstract philosopher, as he and his school undertook a number of dissections to establish the facts about many animals, although many of these texts seem not to have survived [Lennox 2001, chapter 5]. It does seem that Aristotle was not undertaking anything we would consider taxonomy, though, preferring a more teleological or finalist approach to classification of living things, and generally classifying when he did under habitat (land dwelling, water dwelling, and air dwelling) rather than morphology [ably reviewed in detail, with an attempt to reconcile the biological practice with the views expressed in the more technical works, by Charles 2002, chapters 8 and 9].

He was quick to point out a problem with the simple Platonic method of dichotomous classification, although he did not reject the idea of division as such [Pellegrin 1982, 1986]. Many of the categories used in a Platonic diairetic classification were what Aristotle called "privative" categories—defined in terms of what they were *not*, rather than what they were. He proposed instead a method of the decomposition of broader categories into parts on the basis of how the parts differed, but he did not require that each division had to be a dichotomy, as Plato and the Academicians had. There could be many parts in each category. In the

Posterior Analytics [96b15–24], he says, using the term *infimae species* for the most specific division of a topic:[8]

> The authors of a hand-book on a subject that is a generic whole should divide the genus into its first *infimae species*—number e.g., into triad and dyad—and then endeavour to seize their definitions by the method we have described. . . . After that, having established what the category is to which the subaltern genus belongs—quantity or quality, for instance—he should examine the properties 'peculiar' to the species, working through the proximate common differentiae. He should proceed thus because the attributes of the genera compounded of the *infimae species* will be clearly given by the definitions of the species; since the basic element of them all [note: *sc.* genera and species] is the definition, i.e. the simple *infimae species*, and the attributes inhere essentially in the simple *infimae species*, in genera only in virtue of these. [McKeon 1941]

This method came to be known in the Middle Ages as *per genus et differentiam*[9]—"by the general type and the particular difference." Something that was differentiable within a *genus* was known to the Western tradition as a *species*. In Scholastic philosophy, *species* represented a range of things we would now call propositions, sense impressions, and so forth, and this usage persisted through Leibniz to the philosophical discussions of Kant, Mill, and Russell. However, initially a species was merely something that could be differentiated out of a more general concept or term. In the *Topics* [101b16–25], Aristotle defines four "predicables" (that which is predicated of things): definition *(horos)*, property *(idion)*, genus, and accident *(symbebekos)*. Species *(eidos)* is not, in Aristotle's list, a predicable, because it is only true of individuals. This list came to prominence again briefly in the fourteenth-century debates on logic (discussed later).

He extends his discussion in the *Metaphysics* in book Z, chapter 12 [1037b–1038a], by asking what it is that makes *man* (the logical species) a unity instead of a "plurality" such as *animal* and *two-footed*. He argues that the differentiae of a genus can lead to its including species which are polar opposites in their specific differences, but, with respect to the genus itself, there is no differentiation. This makes sense only if each genus is divided further in terms other than the predicates that define the genus. Furthermore, he rejects Plato's dichotomous approach, saying that "it makes in general no difference whether the specification is by many or few differentia, neither does it whether that specification is by a few or by just two." Therefore, he asks whether the genus exists "over and above the specific forms constitutive of it" and answers that it doesn't matter, because the definition is "the account derived just from the differentiae." In the end, we reach the "form and the substance," the last differentia, unless we use accidental features, in which case we will find that we have an incorrect division as evidenced by the differentiae being "equinumerous with the cuts." In short, a species is the form and the substance of the immediately preceding genus, when we reach the last differentia. It is sometimes held that for Aristotle all classification was in terms of absolute definition and essence, and often this is true. But he did allow for an excess or deficiency of some organs or properties of organisms in the *Parts of Animals* [e.g., 646b20], and in the *Rhetoric*, one of the main topics or lines of argument was "the more and the less" [1358a21]. In the eighth book of the *Metaphysics* (Book Iota), he even says, "[I]t is important to understand the kind of differentiation, given that they are the principles of the beings of things. Some things, that is, are marked out by being more or, conversely, less F, by being dense, say, or rare and so forth, which are all instances of the surfeit/deficiency differentiation" [1042b, Lawson-Tancred translation;[10] Aristotle 1998].

Similarly, in the *History of Animals* [486a–486b], he discusses differences being more or less the same property in respect of the genus and, "in short, in the way of excess or defect," "for the more and the less may be represented as excess and defect" [D'Arcy Thompson translation, Barnes 1984]. Species of birds and fish, for instance, may not properly instantiate their genus. The more and the less, however, refer to aspects of the *eidos* that can vary over a range and that can be important for the organism's life [Lennox 2001: 178]. The range is precise and forms part of the differentia of that species.

In *The Parts of Animals*, Aristotle discusses why privative terms are not proper to classification [Book I, chapter 3, 642b–643a]. He says that you cannot properly further divide a privative term, and that precludes it from being a generic (that is, general) term. Worse, a privative classification can include the same group under contradictory terms. Classifications, according to Aristotle, must say something direct and clear. Dividing the world into things that are, and aren't, describable by some predicate is at best only a partial classification, and the taxa that result are not good divisions of the broader genus. Some things that fall under a privative term must be species, but there is no genus out of which those species can be differentiated, and so the privative "genus" is illusory. Plato's mistake was, he thought, to assume that classifications must be made in terms of polar opposites.[11] Aristotle was saying, as we would now describe it, that classification must always be cast in terms of proper sets and subsets. Partial inclusion is not legitimate in a good classification, in effect because it does not make proper sets.

It is traditionally held that Aristotle was inconsistent in the way he used *genos* and *eidos* between the logical and the biological writings [e.g., Mandelbaum 1957], but recent work by Pellegrin, Balme, and Lennox has shown otherwise [Balme 1987a; Lennox 1987, 1993, 1994, 2001; Pellegrin 1986, 1987]. In part, the problem arises because the common view rests mainly on the postmedieval concepts that, we shall see, are derived out of the later neo-Platonic revision of Aristotelian logic. Aristotle is only inconsistent if understood to use the term *eidos* as a technical term in the same way in the logical works as he does in the context of biology, and he doesn't.

Pellegrin says that Aristotle was not aiming to produce a biological taxonomy in *History of Animals*. Instead, he was producing general classifications, and animals happened to be one domain in which he applied that method. What Aristotle treats as genera and species do not answer directly to the modern, post-Linnaean, conceptions of species, although this has sometimes been the default interpretation. We have seen that for him a species is a group that is formed by differentiating a prior group formed by a generic concept. Genera have essential predicates (or definitions), and so do species. Infimae species happen to be indivisible, that's all. In this respect, biological species are no different to any other kind. Pellegrin says, "Aristotle thus conveys by the term *genos* the transmissible type that in our eyes characterises the species, and by *eidos* the model that is actually transmitted in generation. It would be necessary for these two terms to converge and become superimposed for the modern concept of a species to be born. For Aristotle, the [biological] species did not yet exist" [1986: 110].

Pellegrin argues that Aristotle's disagreement with Plato is not that classification by division is wrong, but that one should not proceed by dichotomous division into groups that are defined by a differentia and its contrary. He notes [p. 48], "Although Aristotle condemns dichotomy as used in the Academy, and does so in all the relevant texts, he does not reject division." For Aristotle, he says, the (infimae) species is a group, and is merely the least divisible group, or, in other words, the least inclusive classification. However, in the later logical tradition from which Ray, Linnaeus, and others borrowed their systematic ranks, the smallest *group* category was the genus.[12] *Homo* was a species for them because all present men (women being included) were descendants of a single pair, Adam and Eve. Linnaeus's binomial nomenclature of the genus name and species epithet, as in *Homo sapiens*, was intended, like a personal name, to give the group (the surname, as it were) and a uniquely referring name of that individual parental pair (and their descendants). This distinction must be borne in mind to prevent eisegesis.

Aristotle used the term *universal* to mean any term (predicate) that covered several things. In *On Interpretation* 17a–17b [Edghill translation, Barnes 1984; see also *Prior Analytics*, 24a], he says, "Some things are universal, others individual. By the term 'universal' I mean that which is of such a nature as to be predicated of many subjects, by 'individual' that which is not thus predicated. Thus 'man' is a universal, 'Callias' an individual." A universal need not therefore be something that is literally universally true, but a general term that is predicated of many things. It is important to bear this in mind when discussing the "universals debate" and the question of the individuality of species. A universal is any term that covers two or more subjects.

Aristotle's biological works and those of his student Theophrastus (discussed in the next section) also strongly influenced the later development of biology, and in particular early botany, but one particular doctrine was most influential: the doctrine of the souls of living things in *De Anima*. Soul *(psyche)* here means something like "motivating force"—plants have only a "nutritive soul" [413a21–35, 414a30], animals also have a "sensitive soul" capable of sensation [413b4–9, 414b3], and hence they must have an "appetitive soul," as do all organisms capable of sensation, because they must have some desire [413b21–24, 414b1–15]. Some animals have in addition a "locomotory soul" [413b3, 414b17, 415a7] and one of those, Man, alone also has the power of rational thought, or a "rational soul" [415a7–12]. Soul is the source of movement and growth, and it is the final cause of those faculties [415b9–27], that for which things are generated. This forms the foundation for the later Great Chain of Being tradition, as we shall see, and it was the foundation for the initial classifications and explanations of Cesalpino and Bauhin [Sachs 1890].

Aristotle has thus developed what we might think of as "science by definition." If one can define the proper attributes of a thing and what divides it from other universals, then one has understood and explained it. All those who attempt this method of science prior to the seventeenth century are in effect practicing Aristotle's method. As a result, the terms *genos (genus)* and *eidos (species)* come to have several senses based on the notion of differentiating out the special classes from the general. This has given some historians of biology, particularly those writing from a modern scientific perspective, considerable trouble.

Two issues arise. One is whether Aristotle thought that species were immutable, and the other is whether Aristotle was an essentialist with respect to species. Lennox [1987, 2001] holds that Aristotle did not depend on species being eternal in the biological sense, that they were not fixed. They could come into being and pass away. But the *kinds (genē)* do not come into being and pass away [Lennox 2001: 154]. They are formed by virtue of having the differentia that distinguish them from their superordinate genus. To this extent, then, he is an essentialist and a fixist of kinds. But I do not think it is correct to saddle him with being a species (biological species) fixist. In a telling passage in *Generation of Animals* II.1 [731b24–732a1], he notes that while individuals cannot be eternal, as they are subject to generation and corruption, he admits that if anything can be eternal on earth, it is the *eidos* of men and animals. Note that he admits only the *possibility* rather than the necessity.

In *History of Animals*, Book VIII, chapter 28 [605b22–607a7], Aristotle discusses what will produce variety in animal life—the main cause is locality, including climate. In Africa ("Libya"), "animals of diverse species meet, on account of the rainless climate, at watering-places, and there pair together; and such pairs will breed if they be nearly of the same size and have periods of gestation of the same length" [Barnes 1984]. Zirkle [1959: 640] says that Aristotle insisted that such hybrids were fertile, but it is unclear from the context exactly how different these diverse species were supposed to be. He goes on to say, "Elsewhere also offspring are born to heterogeneous pairs; thus in Cyrene the wolf and the bitch will couple and breed; and the Laconian hound is a cross between the fox and the dog." He then reports that "they" say that the Indian dog is a bitch-tiger third-generation hybrid. If we allow that for Aristotle "species" here does not mean a strict biological species and that travelers' tales are misleading him, he is not so obviously a mutabilist, either. Certainly he is restricting the amount of hybridization here.

As Lennox notes [2001, chapter 5], there is a great gap in practical biology between Aristotle and Theophrastus, and Albertus Magnus's *De Animalibus* in the twelfth century. However, there continued to be philosophical treatments of the idea of division in living things throughout the remainder of the classical period.

THEOPHRASTUS AND NATURAL KINDS

Theophrastus (370–c. 285 BCE) in his botanical work *Enquiry into Plants (Peri phutōn historias)* [Theophrastus 1916] applies Aristotle's notions of classification to the wealth of new specimens being returned to the Greek empire of Alexander from the new conquests. He is the first known author to classify plants in an overall system and may thus be thought of as the first botanical systematist. His translator, Sir Arthur Hort (one of the translators of the first critical edition of the Greek New Testament), notes that Theophrastus's botanical work is guided by a "constant implied question . . . 'what is its *difference?*', 'What is its essential nature?', viz. 'What are the characteristic features in virtue of which a plant may be distinguished from other plants, and which make up its own 'nature' or essential character?" [p. xviii]. In short, Theophrastus is applying Aristotle's notion of classifying into differentia.

He may not have studied all the plants himself. Some descriptions, such as of the cotton plant, banyan tree, cinnamon bush, and so forth, are taken from reports by Alexander's followers who were trained observers. Still, what is most significant about Theophrastus in this context is that his method of classification was an attempt to discover the underlying essence of the kinds of plants based on evidence. His differentiations are almost entirely based on anatomical features of the organisms, although he also allows the habitat (locality) to be employed—for instance between water dwellers and land dwellers. He says, "Now it appears that by a 'part' *[meros]* seeing that it is something which belongs to the plant's characteristic nature, we mean something which is permanent either absolutely or when it has appeared (like those parts of animals which remain for a time undeveloped) permanent" [I.1.ii].

But this is unsatisfactory, because some crucial diagnostic characters, like flowers, inflorescences, leaves, fruit, and so forth, and the shape of the shoot itself, are all temporary or seasonal. He concludes that the characteristics, or "parts," should be chosen from those that are most directly concerned with reproduction [I.2.2]. After setting out the parts he will use (branch, twig, root), he says, "Now, since our study becomes more illuminating if we distinguish different kinds *(eidē)*, it is well to follow this plan where it is possible. The first and most important classes, those which comprise all or nearly all plants, are tree, shrub, under-shrub, herb" [I.3.1], Elsewhere, in VI.1.2, he says "Now let us speak of the wild kinds. Of these there are several classes *(eidē)* which we must distinguish *(diairein)* by the characteristics of each sub-division as well as by those of each class *(genos)* taken as a whole *(tois holois eidesi)*." In the case of the dwarf palm (II.6.10), he distinguishes it as a different *genos* from the ordinary palm.

The translator, Hort, notes in a footnote that Theophrastus "uses *eidos* and *genos* almost indiscriminately. Here *tōn holōn genōn* means the same as *tois holois eidesi*; and below *genōn* and *eidōn both* refer to the smaller divisions called *mere* above." Theophrastus is attempting, in a nonsystematic and perhaps less careful manner, to apply Aristotle's philosophy of classification to botany, but he uses these terms in a more vernacular sense of "form" and "race" or "stock." In any case, his theory of classification, if it can be called that, is morphological and seeks in this the "nature" of the plants. *Species* as such has no special meaning for him here. To back up this interpretation, note that in his *Metaphysics* [Theophrastus 1929], he offers the same view as Aristotle: "Knowledge . . . does not exist without some difference *(diaphoras)*. For if things are other than one another, there is difference; and within universals *(katholou)*, the things that fall under the universals being more than one, these too must differ, whether the universals are genera or species." In particular, he asserts that this is how science is done: "in general it is the task of science *(epistēmēs)* to grasp what is the same in several things, whether it is asserted of them in common and universally or in some special way with regard to each, e.g., of numbers and lines, of plants and animals; complete science is that which includes both these kinds."

Zirkle [1959: 639] says that Theophrastus spent the entirety of Book II showing how species could change from one to another. Given the informal way in which Theophrastus talks about species, we might consider this doubtful. In fact, what he talks about there is generation (including spontaneous generation), "but growth from seed or root would seem most natural" [II.1.1; see also Book I of *De causis plantarum*, 1.2–3, 5.1–2, but he notes that this is sometimes from "unnoticed seeds" in 5.3 and 5.4], and especially vegetative propagation. Zirkle appears to be referring to the "degeneration" of some cultivars into wild forms [e.g., II.2.5]. The oak sometimes "deteriorates from seed" [II.2.6, cf. *De causis plantarum*, I.9.1], for example, so that the child is unlike the parent. Sometimes wild trees like pomegranates, figs, and olives can spontaneously change into a domestic form and vice versa [II.3.1], and so on. It is a very long stretch to call these changes of species. In modern terms, these would either be due to genetic expression of latent varieties, or somatic mutation and development.

One case that is ambiguous is the change from one-seeded wheat and rice-wheat in the third generation when they are bruised (in seed form?) before they are sown [II.3.4], but then he says that "[t]hese changes appear to be due to change of soil and cultivation" and seasonal variation. He does say, as Zirkle noted, that "the water-snake changes into a viper, if the marshes dry up," but it is unclear he thinks this is a species change in our sense. Instead, "when a change of the required character occurs in the climatic conditions, a spontaneous change in the way of growth ensues" [II.4.4]. This is hardly a claim of mutability of species. The remainder of the book consists of a discussion of transplanting, grafting, watering, and planting. He does note that fruit trees are male and female [II.6.6]. Another case is the widely held view that gall wasps come from the seed of the wild fig [II.8.1–2], which causes their fruit to drop, and he also follows Aristotle on the spontaneous generation of animals when the earth is warmed and "qualitatively altered" by the sun [*De causis plantarum* I.5.5]. These are all that I can find that match Zirkle's claim. It will pay to be skeptical of the oft-repeated claim that species were *always* held to be changeable before the seventeenth century.

EPICUREANISM AND THE GENERATIVE CONCEPTION

The Aristotelian and Platonic traditions were not the only ones in the classical period that dealt with species. The atomists, and in particular the Epicurean tradition, also had an account of why forms are as they are. Epicurus's (341–270 BCE) own writings are largely lost, in particular his *On Nature*. However, we have a full account of the Epicurean doctrines in the work *On the Nature of Things* by Lucretius, a first-century BCE Roman disciple of Epicurus. Lucretius tied specific natures of things to the ways in which they came to be:

> If things could be created out of nothing, any kind of things could be produced from any source. In the first place, men could spring from the sea, squamous fish from the ground, and birds could be hatched from the sky; cattle and other farm animals, and every kind of wild beast, would bear young of unpredictable species, and would make their home in cultivated and barren parts without discrimination. Moreover, the same fruits would not invariably grow on the same trees, but would change: any tree could bear any fruit. Seeing that there would be no elements with the capacity to generate each kind of thing, how could creatures constantly have a fixed mother? But, as it is, because all are formed from fixed seeds, each is born and issues out into the shores of light only from a source where the right ultimate particles exist. And this explains why all things cannot be produced from all things: any given thing possesses a distinct creative capacity. [Lucretius 1969: 38, Book I, 155–191]

It is commonly understood that Lucretius gives a more or less faithful exposition in Latin of Epicurus's ideas expressed in Greek some two centuries earlier, in the period following Aristotle of the fourth century BCE [Sedley 2007]. This being so, we can suppose that something resembling the biospecies concept existed by the fourth century BCE. The Epicurean view of species (which is not restricted to biological species—like Aristotle's, it applies to elemental forms of all things but is here illustrated in terms of living things) relies on the potential nature of the composite parts of things. It is, if you will, a kind of *generative conception of species*. He goes on to say that things grow at the right season and are able to live because only then are the right "ultimate particles" (i.e., atoms) available to promote growth. Otherwise, everything could happen, such as children and trees maturing in an instant: "But it is evident that none of these things happens, since in every case growth is a gradual process, as one would expect, from a fixed seed, and, as things grow, they preserve their specific character; so you may be sure that each thing increases its bulk and derives its sustenance from its own special substance" [Lucretius 1969: 38].

We shall see this generative notion of species being struck on repeatedly in the history of the concept both before and after the term *species* attains a technical sense in biology (for example in both Cusa before it and Buffon after). It is of interest that Epicurus's teachings appear to deny Aristotle's acceptance of spontaneous generation [Lennox 2001, chapter 10], given the role spontaneous generation plays in the later history of living species.

Lucretius further expounds the nature of species in the generalized sense of classification of all things in terms of the natures of the atoms that comprise those things, in Book I [pp. 584–598]. Again he appeals to the natures *in potentia* of the constituents as determining the limits of a species.

> Furthermore, since in the case of each species, a fixed limit of growth and the tenure of life has been established, and since the powers of each have been defined by solemn decree in accordance with the ordinances of nature, and since, so far from any species being susceptible of variation, each is so constant that from generation to generation all the variegated birds display on their bodies the distinctive markings of their kind, it is evident that their bodies must consist of unchanging substance. For, if the primary elements of things [i.e., atoms] could be overpowered and changed by any means, it would be impossible to determine what can arise and what cannot, and again by law each thing has its scope restricted and its deeply implanted boundary-stone; and it would be equally impossible for the generations within each species to conform so consistently to the nature, habits, mode of life, and movements of their parents. [Lucretius 1969: 49]

> [E]very species that you see breathing the breath of life has been protected and preserved from the beginning of its existence either by cunning or by courage or by speed. [Lucretius 1969: 191]

Intriguingly, and famously, Lucretius [V, pp. 837–877] and the Epicureans have an "evolutionary myth" of the origins of living species, and in it they suppose that these generative natures were not fixed in the initial period of life. The mixtures of the elemental particles were random, and so all kinds of organisms and monsters were born. Eventually, only those that could propagate remained in existence, and the others died out. It is sometimes held that Lucretius and the Epicureans therefore held a natural selection view of adaptation [the classical locus being Osborn 1894], but in fact they suppose that the species are as they originally were formed by chance and are thereafter kept to the limits of their generative potential. This is not selection as Darwin and Wallace proposed it—there is no variation except in the different but unchanging natures of the characters of the particles that by chance form the species themselves, not within the species.

The Epicureans therefore differed from Aristotle, who held that species were forms that are imposed on the substance of things, instead holding that species are forms generated by the natures of their substances. For Aristotle, material substance is malleable. For Epicureans, it is deterministic of the nature of the things it comprises. Of course, Aristotle, too, held that the four elements he proposed in the *Physics* contribute through the "material cause" to the nature of the objects, but he also allowed for formal, efficient, and final causes. Epicurus and his disciples seem not to allow for any determination of natures other than by the material atoms.

This explains a comment made in Boëthius's *Second Commentary* on Porphyry's *Introduction (Isagoge)* to Aristotle's *Categories* some four centuries later, which is crucial to the transmission of the species problem to the medieval universals debate, and thence to the modern era:

> It is clear . . . that this happened to him [Epicurus], and to others, because they thought, through inexperience in logical argument, that everything they comprehended in reasoning occurred also in things themselves. This is surely a great error; for in reasoning it is not as in numbers. For in numbers whatever has come out in computing the digits correctly, must without doubt also eventuate in the things themselves, so that if by calculation there should happen to be a hundred, there must also be a hundred things subject to that number. But this does not hold equally in argumentation; nor, in fact, is everything which the evolution of words may have discovered held fixed in nature too. [*Second Commentary on Porphyry's Isagoge*, Book I, section 2 (McKeon 1929: 73)]

Boëthius is complaining that the Epicureans and the "others" (other atomists, that is) have presumed that because they have been able to construct a coherent account, that what is said must be true of the things being spoken of. Aristotelians, and the neo-Platonists who followed, held that the physics of Aristotle was based on observable features of the world, while Epicurus's atoms are mere speculations, and hence so are the things that depend on them for their natures, such as species. This is somewhat ironic, given the clearly theoretical nature of the quintessence in Aristotle's cosmology, and even more so given the merely logical role that *genus* and *species* play in Aristotle's categorical logic. Aristotelian essence is, in effect, exactly the kind of reification to which Boëthius, who is defending Aristotle, objects. Although Aristotle's essence concept was not then considered to be an inappropriately abstracted notion [what Whitehead 1938 later called the "Fallacy of Misplaced Concreteness"] when Boëthius wrote, the issue had been raised by Porphyry himself as to whether species exist merely as abstract mental objects (discussed later). The Epicureans were not well regarded during the period of Christian domination [Clark, Foster, and York 2007]. The tradition was held to be an impious and immoral philosophy, by Christian as well as by Jewish thinkers of the postclassical era. The direct influence of Epicurean generative conceptions is thus likely to be sparse before the Enlightenment.

THE HERMETIC TRADITION: SPECIES COME FROM LIKE

Even less immediately influential were the Hermetic writers, although they became significant in the Renaissance. Around 100 CE, a text was written under the pseudonym of Hermes Trismegistus, today known as *Asclepius* I [Scott 1924]. This tradition in part is the direct descendant of Plato and the Academy, with a strong veneer of mysticism and *gnosis*. In this text, nature is the matter that nourishes the forms *(species)* that are imposed on it by God [3c], and in the manner of Plato's souls in *The Republic*, different kinds *(species)* are realized according to the source. The god-kind produces gods, the demon-kind produces demons, and the man-kind produces men [4]. Unlike Plato, however, the author allows that individuals might partake in many kinds, "though all individuals exactly resemble the type of their kind, yet individuals of each kind intermingle *[miscentur]* with all other kinds." A later scribe has interposed the comment that organic bodies receive their kinds by the fiat of the gods, and individual things receive form by the ministration of daemons.

In *Asclepius* II (c. 150–270 CE), matter is considered "ungenerated" [15] but to have a "generative power" and to be creative, reiterated also in Book II (c. 270 CE). The Hermetic tradition is likely influenced strongly by the Stoics. The subsequent influence of the Hermetics was strong in the alchemical and mystery religions, and through them to the gnostic traditions of European thought. It is therefore a minor source for the later Great Chain of Being.

THE CLASSICAL TRADITION OF NATURAL HISTORY

The classical period of biology divides mainly into two" the Peripatetic period deriving from Aristotle and Theophrastus, and the practical or encyclopedic period instituted by Pliny the Elder and Dioscorides, in which botanical and zoological information was assembled for practical use. The latter tradition developed into the herbalists of the Middle Ages, in which medical information was the goal and rationale [Stannard 1968; reprinted in Stannard, Kay, and Stannard 1999]. A later third tradition was the so-called etymological tradition that began with the *Physiologus* (c. 200) and ended with Isidore of Spain's *Etymologiae*.

Pliny the Elder (23–79 CE), who famously perished while observing from a boat the eruption of Vesuvius that buried Pompeii, wrote an encyclopedic account of all animals, birds, trees and so on known to Roman science, called the *Historia naturalis (Natural History)*. It was the standard reference work for some 1,500 years, influencing the later bestiary tradition [Pliny the Elder 1906, 1940–63] and being the foundation for almost all medieval botanical works. In it, he refers to kinds of organisms almost always as "genera"; for instance, he speaks of two "kinds" *(genera)* of camel, the Bactrian and the Arabian [II.xxx],[13] or of kinds of lions [VIII.xx], or six kinds of eagles [X.iii]. There is no emphasis placed on reproduction in Pliny, and most of the descriptions are morphological and behavioral. Likewise, the herbalist tradition that began with Dioscorides' *De materi medica* (first century CE) assumes that there are kinds and sorts, but makes no clear distinction between them. Even so, modern classifications can be fairly clearly mapped onto the species mentioned [Dioscorides 1959: 663–679]. Pliny based much of his material on Theophrastus and Dioscorides [Nordenskiöld 1929: 191].

THE NEO-PLATONISTS: SPECIES AS A PREDICABLE

Aristotelian categories strongly influenced the neo-Platonists, who in turn influenced the medieval scholastics from whom Linnaeus drew his ranking categories. A clear example is the fourth- or fifth-century writer

Martianus (or Felix) Capella. In Martianus's scheme, which is representative of the tradition, genus is the higher grouping and species are the members of the genus. What is a genus with respect to one predicate can be a species with respect to another. Martianus, who was by tradition a farmer in fifth-century CE Africa but more probably a wealthy landowner, did not explicitly deal with the classification of living things and effectively repeated the abstract position of Aristotle's chapter 13 of the *Categories*. His text, whimsically entitled *The Marriage of Philology and Mercury*, was used as a major textbook of the medieval educational program that came to be known as the Quadrivium and the Trivium, for over a thousand years, surely a record for a purpose-written instructional textbook (excluding, perhaps, Euclid). Martianus wrote of a genus being a collection of forms under one name, and species are "man, horse, lion." He wrote that "we also call species forms," which have a "name and definition." "The term and definition of genus are thus determined."[14]

The interesting thing about this is that there is no necessity that a living species should not be further decomposable (see the later discussion regarding Whately's *Logic* of the nineteenth century), somewhat at odds with the standard conception of the Aristotelian metaphysical logic. In the neo-Platonic interpretation of Aristotle, mediated to medieval Christianity by Martianus and Porphyry in the *Isagoge*, via Boëthius in the *Commentaries*, a species was a member of a broader group—a genus—that was formed by a predicate. There was no necessity for any object to be a member of a single genus, and a species might be, with respect to some other predicate, a genus in its own right. In short, species were predicate-relative individuals. However, they were individuals in neither the nominalist sense—name-bearing particulars—nor the Strawsonian sense [Strawson 1964, chapter 8]—historically and geographically delimited objects. They were whatever was differentiable out of the genus. This gave rise to Porphyry's dichotomous notion of classification, although the terminology and many of the concepts were derived from Aristotle's *Posterior Analytics*. Porphyry of Tyre, a student of Plotinus (c. 232 or 234–c. 305 CE) syncretized Plato's dichotomous method with Aristotle's logical division of predicables. Aristotle's conception of the *infimae species* was primarily a matter of logical analysis. Porphyry combined this and Plato's method of classification in the *Sophist* to produce what became known as *Porphyry's comb* or *tree (Arbor Porphyriana)* in the later Middle Ages, which is topologically the same as a cladogram (see figure 2). But a major distinction between a phylogenetic tree and Porphyry's comb is that the former is derived from history, while the latter is derived from diagnosis [Nelson and Platnick 1981].

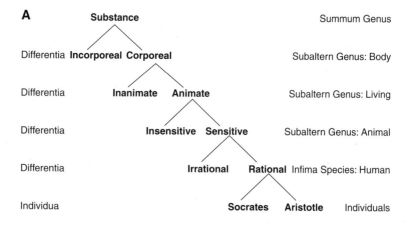

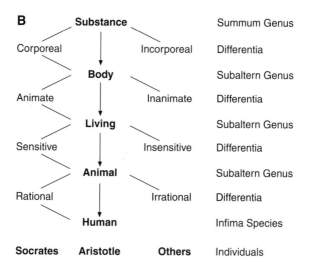

Figure 2 Porphyry's tree. The top version (A) has been adapted from Oldroyd [1983: 29] to make the logical isomorphism with a cladogram clearer; the traditional form of Porphyry's tree is more like the one Oldroyd presents (B), a late example of which [from Baldwin 1901, 2: 714], is shown at the right (C), where "Socrates" has, ironically, been corrupted as "Sortes." See also the discussion by Barnes [Porphyry and Barnes 2003: 109f.], who locates the earliest such tree in the Middle Ages, not in Porphyry's own text. The terms used are those of the late medieval scholastic tradition.

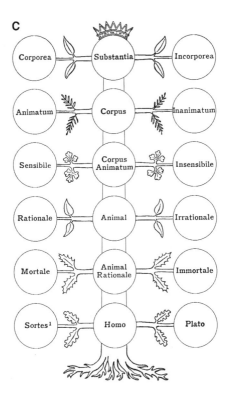

Porphyry treated species slightly differently to Aristotle. In place of the four predicables of Aristotle, Porphyry had five: genus, difference, species, property, and accident [cf. Barnes's commentary §0 in Porphyry and Barnes 2003: 26–32; also see Joseph 1916, chapter 4], replacing *definition* with *species*.[15]

Boëthius reported that Porphyry had raised the issue of whether species and genera exist only in the mind, some 1,400 years before Locke addressed the same issue (discussed later), and indeed well before the nominalists of the fourteenth century:

> As for genera and species, [Porphyry] says, I shall decline for the present to say (1) whether they subsist or are posited in bare [acts of] understanding only, (2) whether, if they subsist, they are corporeal or incorporeal, and (3) whether [they are] separated from sensibles or posited in sensibles and agree with them. For that is a most noble matter, and requires a longer investigation.[16]

This began what we now know as the "Universals" debate and led, fairly directly, to the position that came to be known as nominalism. Why did Porphyry even raise this question? It seems to come out of nowhere. Aristotle had no doubt that the *eidos* was real—it was the form of the thing that existed as a material object. Plato had no such doubt, either—to him *only* Ideas were real.[17] I suspect that Porphyry was responding to the debates over atomism that bridge the period of Aristotle and the Epicureans, and his day, around five hundred years' duration. Plotinus had addressed the arguments of the Gnostics in his *Enneads*, which Porphyry edited. It is possible that the topic had indeed been raised by the Roman Epicureans, who discussed the nature of sensation extensively, and that the Aristotelians in the person of Porphyry are attempting to defend the essentialist account against the atomistic substantist one.[18]

AUGUSTINE: THE MUTABLE IN GOD'S DESIGN

Augustine's *Civitas Dei (The City of God)* was finished in about 426 and espouses a Christianized neo-Platonism of sorts, which is not surprising, as Augustine was in the Roman part of Africa from which neo-Platonism sprang. Although not directly interested in matters of natural history, he did assert in Book VIII, chapter 6, that

> all forms of mutable things, whereby they are what they are (of what nature soever they be) have their origin from none but Him that is true and unchangeable. Consequently, neither the body of this universe, the figures, qualities, motions, and elements, nor the bodies in them from heaven to earth, either vegetative as trees, or sensitive also as beasts, or reasonable also as men, nor those that need no nutriment but subsist by themselves as the angels, can have being but from Him who has only simple being. [Augustine 1962]

Forms (species) are thus maintained by the direct action of God, rather than any internal or innate quality. Augustine's focus is, as the title suggests, on heaven, and bodies of organisms are of interest only so far as they are relevant to resurrection. This lack of interest in the natural world extends until the late Middle Ages, as we shall see. This chapter was influential on Peter Lombard (c. 1100–1160 or 1164 CE) in the much-discussed *Four Books of Sentences* [Book I, Discussion III, chapter 1; McKeon 1929: 190f.] where the same points are made.

One text of Augustine's that has been interpreted to mean he held to an "evolutionary" view is in his commentary on the meaning of Genesis:

> Where, then, were they [*plants, when they were created*]? Were they in the earth in the "reasons" or causes from which they would spring, as all

things already exist in their seeds before they evolve [develop—JSW] in one form or another and grow into their proper kinds in the course of time? . . . it appears [from Scripture—JSW] . . . that the seeds sprang from the crops and trees, and that the crops and trees themselves came forth not from seeds but from the earth. [*De Genesi Ad Litteram (The Literal Meaning of Genesis)*, c. 390 CE, Book V, chapter 4 (Augustine 1982: 151f.)]

However, this is best understood that God created secondary powers that spontaneously generate plants at the right time. This passage does not license an interpretation of Augustine as propounding either fixism or mutablism.

THE MEDIEVAL BRIDGE

BOËTHIUS: THE NATURE OF THE SPECIES IN LOGIC

In his *On Division* (*De divisione*, c. 505 CE), Boëthius (480–524 or 526 CE) discussed the nature of classification by division. In an extended introduction to the topic, he sets out influentially the basis for the "classificatory logic" of the next 1,500 years, and as he wrote in Latin, this was almost as influential in the development of Western thought as Augustine's works were. Nearly all his examples are based on animal/ human and similar biological cases, but it should not be thought this is a biological concept, any more than Aristotle's.

He distinguishes division of genus into species from whole into part and utterances into "proper signification." His version of the *diairesis* is not binary, however. "Every division of a genus into its species has to be made into two or more parts, but there cannot be infinitely many or fewer than two parts" [877C, Kretzmann and Stump 1988: 14]. The matter of accidents and essences applies here to utterances (that is, of the meanings of propositions), not species, and he notes that while a genus is the whole of which a species is a part, more universal utterances such as equivocation are not wholes in nature [878D–879A].

NOMINALISM: SPECIES ARE IN THE UNDERSTANDING

There was little advancement on biological species concepts in the medieval period, which is usually delimited from 500 to 1500 CE. This is not because the medievals were unimportant—very far from it. The revival of the universals debate in the late eleventh century, by Roscelin, Abelard, and eventually the nominalist school, was critical in bringing the notion of genera and species to the forefront of Western thinking [Leff 1958]. Once there the idea was taken up by the nascent biological sciences in the seventeenth century, but so far as I can tell, nominalism did not directly influence biological practice, although it did influence Locke.

The nominalists held that universals existed solely in the understanding, or at least that is the nominalist position as later understood. Probably Roscelin did not assert this but merely rejected the view known as Realism (that universals are real things that exist apart from the substance of the things that are within them), and Ockham's version might be better called Conceptualism rather than an outright denial of universals.

The conception of species in this debate centered largely on the objects of knowledge rather than on anything particularly biological. However, as evidenced in Book IV of John Scotus Eriugena's *De divisione naturae (The Division of Nature)* of the ninth century, biological examples are used to illustrate the discussions [McKeon 1929: 107–141]. Peter Abelard (or Abelaird, 1079–1142) was influenced by the nominalist claim, and studied under Roscelin. He convinced William of Champeaux, his lecturer, to modify his Realist position [McKeon 1929]. In his *Gloss on Porphyry*, he asserted that universals are names only, akin to proper nouns but applicable to many things rather than one thing, based on a common likeness. The universal noun does not refer to a thing that is universal, for only individuals exist. His views led to his being declared heretical by later writers, including Bernard of Clairvaux. For him, *species* is a purely logical notion, and he says, "[C]onsidering the nature of species in man, I find at once from the nature of the species the argument for proving animal" [p. 211]. That is, if you understand what sort a man is, then you know that it includes the species Animal, just as Aristotle's *De anima* asserted, and this is a logical truth.

William of Ockham (or Occam, c. 1300–1349) is perhaps the most enduringly influential of the medieval nominalists. For him, logical species are just a way to recollect similar individuals already encountered, and a general term cannot be abstracted from a *single* individual, but only a number of individuals encountered, as he says in *The Seven Quodlibita*, Quodlibitum I, question 13 [in McKeon 1930: 365]. Hence, to assert that something coming from a distance is an animal, one must already have that concept by recollecting prior individual animals. This demonstrates that the nominalist is attempting a "bottom-up" form of classification, based on observed cases. Moreover, species, logical or otherwise, are things that have the same "power" [Quodlibitum V, question 2], but concepts of them are of the "second intention" (that is, have been categorized by the mind):

> [T]hat concept is called a second intention which signifies precisely intentions naturally significative, or which sort are *genus, species, difference*, and others of this sort [i.e., heads of predicables] for as the concept of man is predicated of all men . . . , so too one common concept, which is the second intention, is predicated of first intentions [a conception of a thing formed by the first or direct application of the mind to the individual object; an idea or image] which stand for things, as in saying, *man is a species, ass is a species, whiteness is a species, animal is a genus, body is a genus*; in the manner in which *name* is predicated of different names . . . and this second intention thus signifies first intentions naturally, and can stand for them in a proposition, just as the first intention signifies external things naturally. [Quodlibitum IV, question 19]

For Ockham, the general term is the name of a number of concepts formed from what might later be called "sense impressions." To say something is a universal is to say it is the name of a number of individual names, each of which "stands for" the "external" (to the mind) individual. The second intention term is thus not a universal in itself, but neither is it an individual; it's just a name.

Another thing that occurred in the period was the advancement of species to a predicable. In the later Middle Ages, Porphyry's tetrad—genus, difference, species, property, and accident—was much debated [Green-Pedersen 1984: 118–121], and it was not exactly followed by all, as there was debate over whether species should be considered "predicate-types," because, as Green-Pedersen says, the species "can only be predicated about individuals, and there can be no science about individuals," a view held consistently through the Middle Ages [p. 120]. In the fourteenth (and increasingly in the fifteenth) century, though, the suggestion is made in the commentaries that the species is an addition *(annexum)* to the genus. This, in effect, would make *species* into a class concept like *genus*, which it was not before. Abelard, as early as the twelfth century, seems to accept the Porphyrian/Boëthian taxonomy of classes as including species.

THE HERBALS AND THE BESTIARIES:
MEANING AND MORAL SPECIES

In the later Middle Ages, particularly in the two centuries leading up to the modern scientific era, bestiaries and herbals developed, acting as precursors to biological classification [Arber 1938]. Herbals were repositories of useful pharmacological knowledge, arranged alphabetically most of the time, and derived from the tradition begun in the first century CE by Pliny the Elder and his contemporary Dioscorides, who assembled such information in a work entitled *De material medica*. The herbal traditions developed in an attempt to map the classical names of plant species onto known colloquial names, in order to ensure that medical preparations were reliable [Stannard et al. 1999, chapters 1 and 2]. The earliest herbal is by Crataeus, physician to Mithridates VI, king of Pontus (111–64 BCE), but only fragments survive in Dioscorides' work. Herbals underwent a decline in popularity in the later Middle Ages and a resurgence with the development of printing.

Bestiaries were usually a combination of moral tales, in which animals represented the virtues or vices, and exotic geography books [McCulloch 1962]. Typically, they relied on a Greek composition of around 200 CE, entitled the *Physiologus* [Curley 1979], which remained widely read until the late Middle Ages and was translated into Latin in the fourth century, French in the thirteenth century, and Middle English shortly before that [Wirtjes 1991]. The work was a deliberately theological Christian work, in which each animal illustrated vices or virtues.

Another source was Isidore of Seville's [2005] *Etymologiae* (c. 630), which purported to give the linguistic origins of the names of beasts and all other matters, to make a theological point [Stresemann 1975: 8]. As Wirtjes, the translator of the Middle English edition, says [pp. lxviii–lxxix], bestiaries were intended to show Nature as a second Book of God alongside the Bible, to show the Christian how to live morally. As such, they do not express much of a view about the nature of species. A medieval bestiary, known in fact as *The* Bestiary [White 1954], simply refers to "kindes," and Hugh of Fouilloy's (1111?–1172?; probably not the author, anyway) *The Aviary (Aviarum)*, written between 1132 and 1152 refers to "species" of hawks: "There are two forms *[species]* of the hawk, namely, the tame and the wild." Clearly, there is no generative or biological conception in play here. Species are just sorts of things. In a contemporary work, though, *The Book of Beasts*, that was itself based on the ever-present *Physiologus* and Pliny [Pliny the Elder 1940–1963] via the work of Solinus, the author several times *does* exhibit the notion of "species" as being a kind of beast that breeds true, such as stags [p. 39], birds in general [p. 107], and shellfish [p. 214]. However, he also includes hyenas as a kind of animal, even though it appears he thought of the animal as a hybrid.[1] In the century following the early twelfth century, colloquial bestiaries became common and popular.

There are two major, and a number of minor, exceptions to the moral form of bestiaries in this period. Minor exceptions include Thomas de Campitré (1210?–1293) in his compilation entitled *De natura rerum* and Konrad von Megenberg (1309?–1374) in a colloquial text *Puch der Natur (Book of Nature)* [Stresemann 1975: 8]. The two major exceptions are the emperor Frederick II and his acquaintance Albertus Magnus.

FREDERICK II: THE HERETIC FALCONER

Frederick II of Hohenstaufen (1194–1250) was, to say the least, an interesting man [Wilkins 2005]. Excommunicated twice, although he was the Holy Roman Emperor (and king of Sicily and Jerusalem), for challenging the supremacy of the pope over secular power, he nevertheless deviated from the Norman tradition and installed a court of cultural and scientific sophistication.[2] He was greatly taken by the recent translations of Aristotle and Averroës (Ibn Rushd, 1126–1198) by Michael Scot (1175–1235) of various natural history texts, including the *Liber animalium (De animalibus historia, De partibus animalium, De generatione animalium)* of Aristotle. He wrote *De arte venandi cum avibus (The Art of Hunting with Birds)*, published around 1248, and is reputed to have replied to one of the letters of the Mongol Khans demanding that he submit to the Khan's power, that he would gladly resign his throne if he were allowed to be the Khan's falconer [Abulafia 1988: 267]. Records of the period from 1239 to 1240 in his court indicate that falconry took a close second place only to the affairs of government. Inspired by Aristotle's *De animalibus*, Frederick had a passion for falconry, recently adopted by the European aristocracy from the Arabs, but he was disappointed in Aristotle's lack of accuracy. He notes in the preface to *De arte*:

> Inter alia, we discovered by hard-won experience that the deductions of Aristotle, whom we followed when they appealed to our reason, were not entirely to be relied upon, more particularly in his descriptions of the characters of certain birds.
>   There is another reason why we do not follow implicitly the Prince of Philosophers: he was ignorant of the practice of falconry—an art which to us has ever been a pleasing occupation, and with the details of which we

are well acquainted. In his work "Liber Animalium" we find many quotations from other authors whose statements he did not verify and who, in their turn, were not speaking from experience. Entire conviction of the truth never follows mere hearsay. [Wood and Fyfe 1943: 3f.]

This statement is revolutionary and marks the beginning of the end of the Middle Ages mind-set toward natural history, albeit for the purposes of a sport. He also says, "Our purpose is to present the facts as we find them." Unfortunately, due to the author's heterodoxy, this view was not directly widely influential, apart from influencing Albertus Magnus. Still, it does represent an attention to the actual facts of observation of organisms, which probably meant that such things as breeding compatibility between types were beginning to take priority over form, or arbitrary distinctions such as the one in [Pseudo-]Hugh's *Aviarum*.

In Book I of *De arte venandi*, Frederick adopts Aristotle's division of birds in *On Animals* into waterfowl, "whose organs are so fashioned that they may remain for indefinite periods immersed in water," land birds, and "neutral" birds that may live in either habitat [chapter 2], because it suits the usage of falconry experts, and then notes that "they may also be divided into various genera and these again into a number of species" [p. 7].

He notes in chapter 3 that birds "may be classified in still another manner—as raptorial and nonraptorial species." In the Bologna manuscript, he notes that the raptorial birds "are the eagles, hawks, owls, falcons and other similar genera." But in the Vatican Codex, the paragraph reads thus: "It was the habit of Aristotle and the philosophers to classify objects into positive and negative groups and to begin their discussions with the positive. Since it is our purpose to give special attention to raptorials, we shall first consider the nonrapacious (or negative) varieties; afterward we shall consider at length raptorial birds." Since this is in fact opposed to Aristotle's method of classification, in as much as it allows for privative groups, this paragraph is perhaps more based on the medieval understanding of logic from Boëthius than directly on Aristotle's own works, and may be considered spurious. Possibly the distinction had not become popularly (among educated nonscholars) recognized.

Later in that chapter, he promises a treatise on the "genera into which raptores [sic] are divided, and the species in each genus," which was never produced. But he does note that "the same genera and species are given different names by diverse authors. Sometimes the same bird may have a variety of synonyms; and the same name applied to diverse birds so dissimilar that one cannot establish the true identity of a species simply by its name." He therefore notes, in complete anticipation of later tax-onomic problems, that

> a description of the essential characters of individual birds [i.e., of a species] is more difficult to furnish, whether they resemble or are different from another in the shape of the limbs, the movements they make, the way they feed, the care of their young, their mode of flight, and their style of defense. Let it, however, be remembered that, in general, their bodily conditions and their other peculiarities are due to definite causes. [p. 10]

This is an amazing statement for the thirteenth century. He allows not only that species must vary in their traits but that they are *caused* to vary. He goes on to say that different localities will have different genera and species, or a location may be the only habitat of species not found else-where. He even says that a genus might be found in many localities but with "a different color, or varying in other respects." It is not entirely exaggeration when Stresemann says of him that no direct observer among ornithologists until Konrad Lorenz in 1933 surpassed him in "variety of experience and acuteness of interpretation" [Stresemann 1975: 11].

The way he refers to species is so clearly in line with modern usage that he might be considered to have been the first to give a truly biological account. This is in part due to the fact that he had practical concerns—breeding birds. When discussing bird reproduction, in chapter 13-E, he notes, "Nature in her endeavor to preserve the race by the continuous multiplication of individuals has decreed that every species of the animal kingdom, whether it progresses by the use of wings or walks on the ground, shall take pleasure in sexual union so that they may seek instinctively to bring about such enjoyment" [p. 49]. Hence also, members of species know "instinctively" how to build nests of a "special design" and raise their children.

His empirical bent is further exhibited when he discusses the old myth that the barnacle goose arises by spontaneous generation from dead wood, a view still held in the sixteenth century by Julius Scaliger [Amundson 2005: 37, from Hacking 1983: 70]. Frederick says:

> There is also a small species known as the barnacle goose, arrayed in motley plumage . . . , of whose nesting haunts we have no certain knowl-edge. There is, however, a curious popular tradition that they spring from dead trees. It is said that in the far north old ships are to be found in whose rotting hulls a worm is born that develops into the barnacle goose. This goose hangs from the dead wood by its beak until it is old and strong enough to fly. We have made prolonged research into the origin and truth of this legend and even sent special envoys to the North with orders to bring back specimens of those mythical timbers for our inspection. When

we examined them we did observe shell-like formations clinging to the rotten wood, but these bore no resemblance to any avian body. We therefore doubt the truth of this legend in the absence of corroborating evidence. In our opinion this superstition arose from the fact that barnacle geese breed in such remote latitudes that men, in ignorance of their real nesting place, invented this explanation. [p. 51f.]

As it happens, *Branta leucopsis* breeds in the northern parts of Europe, in hilltops, cliffs, slopes, and islands with nearby coasts or rivers with plenty of grass and other vegetation to graze. It spends winters in salt marshes, lowland fields near coasts, and offshore islands with suitable grassland.

So, Frederick is inclined to think that species are caused by generation from parents via sex, rather than accept the view of spontaneous generation for some, a view held well into the eighteenth century [Farley 1977]. Furthermore, in chapter 23-I, he notes that "productive Nature" formed organs for each species that are benevolent for one species but malevolent for another, and that it

> must be held, then, that for each species, and each individual of the species, Nature has provided and made, of convenient, suitable, material, organs adapted to individual requirements. By means of these organs the individual has perfected the functions needful for himself. It follows, also, that each individual, in accordance with the particular form of his organs and the characteristics inherent in them, seeks to perform by means of each organ whatever task is most suitable to the form of that organ. [p. 57]

All that is missing here is a claim that the most fit will become the most widespread form, and we would have an anticipation of Darwin. However, he regards identification of species to be a matter of recognition of plumage; and at one point [p. 122], he disagrees with those who reject a particular variety of peregrine based on its plumage as being a true peregrine, showing that identification marks were insufficient to act as the definiens or essence of a species.

ALBERTUS MAGNUS ON BEASTS AND PLANTS

Albert of Lauingen, later known as "The Great," is better known as Albertus Magnus (1193–1280). He was a Dominican friar who, among other things, taught Thomas Aquinas and was among the first to comment on the works of Aristotle that were coming out of the Arab tradition. Hitherto, much logical discussion in the European intellectual world was founded on Peter of Lombard's *Sentences* and the books of Boëthius.

Now, works like Aristotle's *De motibus animalium*, which Albert found in Italy, and *Historia animalium* became commonplace.

Albert also had a youthful interest in falconry, which was the proximate source of the love of nature evidenced in several medieval authors, including, as we have seen, Frederick II [Albertus Magnus 1987: 4, 19; Stresemann 1975, chapter 1]. Unlike Frederick, though, he was and remained an orthodox member of the church, later being given the title "Doctor Universalis," and his ideas were taught in subsequent educational institutions. He was directly familiar with Frederick's work on falconry. Albert may also have learned personally from Frederick's falconers, to whom the emperor had given him access.

His major work on natural history is *De animalibus*, books 22 to 26 of which cover beasts and man [Albertus Magnus 1987]. He also produced a long text, based more on his own observations than *De animalibus*, on plants, *De vegetabilibus*. In both, he relied a lot on Pliny, but occasionally he corrects some of the more mythological accounts of Pliny. Nevertheless, the Chimera, the Manticore, and other mythological beasts appear in *De animalibus*, along with some acute and accurate observations, recounted in James Scanlon's introduction to the edition and translation cited here. Occasionally, he gave several separate descriptions for the same animal (e.g., *Alces alces*, the European elk, being named as "Alches," "Aloy," and "Equicervus").

In *De animalibus*, books 22 to 26, Albert describes, species by species, various animals known, after a short tract on man, "the most perfect of all animals," reflecting the Great Chain consensus of the period. In this tract in book 22, he notes that human reproduction is due to sexual intercourse "by which the potentialities of the two sexes are inextricably mingled" [p. 59]. He cites *De coito* of Consantine of Cassino (Constantinus Africanus, c. 1010–1087, a Dominican monk who translated Arabic medical works), who, he said, held that "the Creator unmistakenly wanted the animal kingdom to endure as a stable entity, never to die out. To this end He ordained the class of animals to be continually renewed by the coupling of the sexes and reproduction of the species so that none would be lost."

In tract 2, he begins with the quadrupeds, since these are the class of animals from which the most significant domesticated animals come, foreshadowing Buffon's arrangement six centuries later. The internal order is alphabetically arranged in Latin, beginning with "Alches" (Elk). His classifications also include whether the animals are "wet" or "dry" in the Galenic scheme. However, he allows for a higher genus than the biological species level itself—for example, where when discussing the "Cefusa" mentioned in Solinus's *Collecta rerum memorabilium* (one of his major sources, who also collected much from Pliny's *Natural history*), possibly a gorilla, he assigns it to "the genus of simians" [p. 93]. This is not a precursor to Linnaean binomials, though. Each species is briefly described, except for the oddities (the elephant, for instance), common animals (the hedgehog, the wolf), and domesticated animals. The entry on horses runs for forty-one manuscript pages and includes details on care, diseases, and so forth. In cases of hybrids, such as the Mule [*Mulus*, p. 160], Albert notes that it is sterile because it is "produced from male and females sperms that are quite dissimilar in nature," although he still regards it as a species. He correctly notes that female mules may occasionally give birth to viable young but ascribes the cause to the heat of the country countering the internal cold of the animal. He says of apes *(Symia)* that the animal "admits of many species." He includes a monkey as an ape, not making a distinction between tailed and nontailed simians.

In book 23, he describes bird species, which, although they are animals, are different to the "general sense" of that term [p. 188]. Again, the coverage he gives varies according to their oddity, commonplace nature, or domestication, with falcons and hawks given extensive coverage, following Frederick. He notes that different species of falcon have different-colored feet and plumage [p. 224]. He describes ten species of "noble falcons," three "inferior falcons," and four species of hybrids [chapter 16], noting that while only four crossbred species are known, there are probably more, and also in other groups of species such as goshawks, sparrow hawks, and eagles [p. 247]. Oddly, he includes bats *(Verspertilio)* among the birds, presumably because they fly, although he notes that they look like winged mice. That this is a purely arbitrary Aristotelian system based on overall habitat is confirmed by the inclusion of crocodiles, hippopotamuses, seals, dolphins, and whales among the fishes in the book on aquatic animals [book 24]. It is clear that for Albertus, *species* is a kind term that relates shared properties but that is maintained by sexual reproduction. We may take this as the best of late medieval opinion.

It is interesting that Albert also addresses the barnacle goose question, noting that he and his friends had bred one with a domestic goose, and that the spontaneous generation account is "altogether absurd as I and many of my friends have seen them pair and lay eggs and hatch chicks" [Raven 1953: 67f., quoting book 23, 19]. Raven documents many of his correct observations, and some of his repeating falsehoods, and Stannard

[1979] shows that of all the plant species Albert lists, nearly all can be identified with a modern species. Nevertheless, Albert did list in book 7, 54–64, five ways plant species could change their species: by improvement or deterioration of seed, by being cut down and shoots being of another species, when cuttings grow as vines rather than oaks ("in Alvernia"), when a tree grows rotten and springs forth a different plant, and by grafting. Amundson [2005: 36] considers this a kind of species mutability, but I suspect it is due more to the fact that in cultivation, "species" is not the same thing as a *biological* species, but rather a synonym for *kind*. As Stannard notes [1980: 366n.], "In medieval nomenclature, *genus, species, varietas,* and *forma* were used interchangeably. Sometimes Albert uses *genus* to denote what we would accept as a *genus* . . . but in other places his *genus* is practicably equivalent to our *species*."

The persistence of the prescientific tradition of the herbals, and of spontaneous generation, has something to do with the fact that as a heretic, Frederick's work was not a safe source for later workers; indeed, later writers accused him of atheism and heresy on several counts. But since Albert's work *was* used, and indeed he was canonized, it becomes hard to account for this purely in those terms. Rather, it seems that what stopped Frederick, Albert, and others such as Roger Bacon from influencing natural history and inaugurating an early empiricism was the fact that natural history was not seen as an end in itself until much later, but rather as an adjunct to theology and homiletics. That said, it is worth noting that a scientific attitude was not beyond the reach of a medieval educated man like Frederick, and that acuity of observation, when driven by immediate interests for which accurate information was required, was something that could be attained in a way that is quite modern. The shift to factual science can thus be seen to be driven by practical matters—in this case, falconry. However, there was a decline in empirical work on animals and plants in the fourteenth century and for a century after.

THOMAS: SPECIES AS INDIVIDUALS

Thomas Aquinas's discussion in the *Summa* [Book I, question 86] does not materially advance the matter and can be treated for our purposes here as straight transmission of the prior Scholastic logic. Aquinas treats it as a question of knowledge, and the "species" he considers is that of *species intelligibilus* [answer 1]:

> Our intellect cannot know the singular in material things directly and primarily. The reason of this is that the principle of singularity in material

things is individual matter, whereas our intellect, as I have said above, understands by abstracting the intelligible species from such matter. Now what is abstracted from individual matter is the universal. Hence our intellect knows directly the universal only.

When he does discuss the more general sense of species, he repeats the standard view, although he makes an interesting observation about infimae species being individuals: "These [species infimae or specialissimae] are called individuals, in so far as they are not further divisible formally. Individuals however are called particulars in so far as they are not further divisible neither materially nor formally" [In lib. X Met., lecture 10, 2123, quoted in McKeon 1930: 498].

So there are two senses of "individual" in play here, says Thomas. One is simply that it is atomic—not further formally divisible. But there are also individuals that are not *materially* divisible, and they are particulars. For Thomas the infimae species is a formal, but not a material, individual. Of course, the question of whether species are individuals and what kind of individual they might be is a vexed modern issue, as we shall see.

Thomas allows that living species can be generated by spontaneous generation from putrefaction [Summa I. 73.1, objection 3]. However, he says:

Species, also, that are new, if any such appear, existed beforehand in various active powers; so that animals, and perhaps even new species of animals, are produced by putrefaction by the power which the stars and elements received at the beginning. Again, animals of new kinds arise occasionally from the connection of individuals belonging to different species, as the mule is the offspring of an ass and a mare; but even these existed previously in their causes, in the works of the six days. . . . Hence it is written (Eccles. 1:10), "Nothing under the sun is new, for it hath already gone before, in the ages that were before us." [reply to objection 3][3]

In other words, species only express what potential they have already been given by God at creation. This hardly licenses Zirkle's claim that Aquinas held a mutabilist view [Zirkle 1959: 640]. Also see his De principiis naturae, in which he restates this view of potentials expressed in generation.

SPECIES AND THE BIRTH OF
MODERN SCIENCE

NICHOLAS OF CUSA: CONTRACTED SPECIES

Nicholas of Cusa (1401–1464) in his *On Learned Ignorance* [c. 1440; cf. Hopkins 1981] represents a bridge between the medievals, who were rediscovering Aristotle but using the categories bequeathed to them by the neo-Platonists, and the Renaissance era. Cusa was an eclectic, and his comments on categories show an influence from Pythagorean as well as neo-Platonic sources. He was also an early proponent of the universal character or language that we will consider later [Rossi 2000]. He held, for example, that ten is the supreme number and that all unity is found in it [Book II, chapter 6, §123]. He thus justifies the Aristotelian ten *topoi* (topics) and says:

> [T]he universe is contracted in each particular through three grades. Therefore, the universe is, as it were, all of the ten categories *[generalissima]*, then the genera, and then the species. And so, these are universal according to their respective degrees; they exist with degrees and prior, by a certain order of nature, to the thing which actually contracts them. And since the universe is contracted, it is not found except as unfolded in genera; and genera are found only in species. [§124)]

He proposes that the universe is not discretely differentiated but that species exist, though they vary by degrees. Cusa is naturally aware of the nominalist debate and takes a pretty standard Aristotelian view of the matter: "But individuals exist actually; in them all things exist contractedly.

Through these considerations we see that universals exist actually only in a contracted manner. And in this way the Peripatetics [Aristotelians] speak the truth [when they say that] universals do not actually exist independently of things. For only what is particular exists actually. In the particular, universals are contractedly the particular" [§124]. In effect, he is saying that species are particulars (individuals), ignoring the fourteenth-century discussions. All general things, such as universals, and indeed the entire universe, only actually exist "in a contracted way" [§125], as particulars, although things of the same species share in a specific nature:

> For example, dogs and the other animals of the same species are united by virtue of the common specific nature which is in them. This nature would be contracted in them even if Plato's intellect had not, from a comparison of likenesses, formed for itself a species. Therefore, with respect to its own operation, understanding follows being and living; for [merely] through its own operation understanding can bestow neither being nor living nor understanding. [§126]

Species are real, independent of the mind, and living species are caused by their "common specific nature." Cusa thus answers Porphyry's question: the understanding gathers species and genera together through comparison, so that these universals are likenesses of nature. Genera and species exist both in the mind and in nature. "Therefore, in understanding, it unfolds, by resembling signs and characters, a certain resembling world, which is contracted in it" [§126].[1]

Later, in Book III, Cusa defines the universe as existing "contractedly in plurality," unlike God, who is "the Oneness of the Maximum" existing "absolutely in itself" [§182].

> Now, the many things in which the universe is actually contracted cannot at all agree in supreme equality; for then they would cease being many. Therefore, it is necessary that all things differ from one another—either (1) in genus, species, and number or (2) in species and number or (3) in number—so that each thing exists in its own number, weight, and measure. Hence, all things are distinguished from one another by degrees, so that no thing coincides with another. Accordingly, no contracted thing can participate precisely in the degree of contraction of another thing, so that, necessarily, any given thing is comparatively greater or lesser than any other given thing. Therefore, all contracted things exist between a maximum and a minimum, so that there can be posited a greater and a lesser degree of contraction than [that of] any given thing. [p. 124]

No individual member of a species, since it would be a contracted thing, can therefore exhibit or instantiate all the features of the species, and so there is in the actual organisms of a species (or members of any nonbiological species) variation in the degree to which they participate in the specific essence. According to Cusa, then, there is variation both within and between taxa. Here we see the beginnings of the "type" concept: while the type itself is definable in terms of some necessary and sufficient conditions, "members" of the type can diverge from it or not instantiate it fully. The only limit "of species, of genera, or of the universe" is "the Center, the Circumference, and the Union of all things." Here we see also the underlying assumption of the Great Chain of Being. He holds that each genus has a species that is "highest," which is coincidental with the lowest species of the superordinate genus, "so that there is one continuous and perfect universe." Since the "maximum union" is God, species of lower and higher genera are not united in something that cannot vary, but in a "third species" in which individuals differ by degrees. He relates this to the *scala naturae* explicitly, in "the books of the philosophers" (that is, in Aristotle and his successors). He says:

> Therefore, no species descends to the point that it is the minimum species of some genus, for before it reaches the minimum it is changed into another species; and a similar thing holds true of the [would-be] maximum species, which is changed into another species before it becomes a maximum species. . . . Accordingly, it is evident that species are like a number series which progresses sequentially and which, necessarily, is finite, so that there is order, harmony, and proportion in diversity. [§§ 185–187]

The gradualism of the Great Chain is evident, as also is the influence of the neo-Platonic view of classification. Of particular note is that Cusa's examples are biological ones. In fact, Cusa apparently suggested (but did not do) experimental work on plants to uncover their natures [Morton 1981: 98f.]. More importantly, he allows for a formal and gradual change from one species of a higher level to a species of the next level below. This is not in any sense evolutionary, but it lays some groundwork for an evolutionary account to develop later.

MARSILIO FICINO: THE PRIMUM OF THE GENUS

Marsilio Ficino (1433–1499) was responsible for a number of neo-Platonic texts being translated and published under Cosimo de Medici. Among other texts, he oversaw texts by Porphyry, Proclus, Plotinus, and Dionysius the Areopagite; all source texts for the neo-Platonic philosophy [Cassirer, Kristeller, and Randall 1948: 185]. In his discussion of the genus—species distinction, he is at pains to view the logical progression of Aristotle as being a description of the actual (but not temporal!) progression of things, and God, of course, is the source of all things. In his *Five Questions Concerning the Mind*, Ficino writes [Cassirer et al. 1948: 194], "The motion of each of all the natural species proceeds according to a certain principle. Different species are moved in different ways, and each species always preserves the same course in its motion so that it proceeds from this place to that place and, in turn, recedes from the latter to the former, in a certain most harmonious manner." Here we have again the generative notion of species that we saw in Lucretius. However, unlike the materialist account of Lucretius, Ficino's view is based on the "principle" of the species and of the *primum* of a genus (the ontological principle *primum in aliquo genere*). Each genus has what later came to be thought of as a *type species*, a first and highest example of the kind, according to Ficino [Cassirer et al. 1948: 189], the *primum*, which is the species that is purely of the genus, with no defining essences other than those of the genus. God is, of course, the *primum* of the genus Being, and from him all being flows, in the standard neo-Platonic way. Things do what they do because they share the limits of their species and are constrained by the end of that genus. The *primum* does it best of all.

With elements, plants, and "brutes," Ficino gives the Aristotelian accounts—the elements have heaviness, and so they fall. Plants and animals have a nutritional and generative power (compare Aristotle's nutritive faculty, *De Anima*, 413a20–33, and the discussion of reproduction in *De generatione animalium*). Species are perpetuated because they have an end, or rather, because to be a species is to have an end. Ficino presents this as a prelude to discussions of the mind that do not concern us here.

THE GREAT CHAIN OF BEING

That the philosophical view deriving from the neo-Platonists had reached popular culture is evidenced in Alexander Pope's *Essay on Man* (1733), Section VII:

> Vast chain of being! Which from God began;
> Natures ethereal, human, angel, man,
> Beast, bird, fish, insect, who no eye can see,
> No glass can reach; from infinite to thee;
> From thee to nothing.—On superior powers

Were we to press, inferior might on ours;
Or in the full creation leave a void,
Where, one step broken, the great scale's destroy'd:
From Nature's chain whatever link you like,
Tenth, or ten thousandth, breaks the chain alike.

This view is known as the *Great Chain of Being* [Lovejoy 1936], and it has a history that arises from Aristotelian concepts, through the neo-Platonists, into the Middle Ages and the revival of Aristotle in the fourteenth through to the sixteenth centuries [Kuntz and Kuntz 1988]. The predicables that defined organic species were an ascending scale of increasingly "perfect" features: from being, to growth, to animation, to rationality. Raymond Lull is perhaps the exemplar of this view [Rossi 2000]. In his view, around 1512, the Chain of Being was a series of steps in a staircase to heaven.

The Great Chain view consisted of a number of related theses held in varying ways by its adherents. One of these, named by Lovejoy [1936: 52] *the principle of plenitude*, has it that the world is as full of all the things it could be, or, as Lovejoy himself stated it:

> [T]he universe is a *plenum formarum* in which the range of conceivable diversity of *kinds* of living things is exhaustively exemplified, but also any other deductions from the assumption that no genuine potentiality of being can remain unfulfilled, that the extent and abundance of the creation must be as great as the possibility of existence and commensurate with the productive capacity of a "perfect" and inexhaustible Source, and that the world is better, the more things it contains.

In short, everything that can be, is, and the world is made to be everything it can be. This is where Leibniz's doctrine of the *lex completio* came from, that Voltaire so wickedly caricatured in his *Candide* as the teachings of Dr. Pangloss. It is found in Plato's writings, but not in Aristotle, who famously wrote in the *Metaphysics* [II, 1003a 2; and XI, 1071b 13] that "it is not necessary that everything that is possible should exist in actuality," and "it is possible for that which has a potency not to realize it" [quoted in Lovejoy 1936: 55]. However, the second plank of the Great Chain is the law of continuity (Leibniz calls it the *lex continui*)—that all qualities must be continuous, not discrete. While Aristotle did not make all things linear, arranged in a single ascending series, he did require that there be no sudden "jumps," from which the medieval claim *natura non facit saltus* (nature does not make leaps) came, which we shall meet again. Aristotle's version did not itself insist that one would classify a single living being in one and only one series, nor that an organism that is graded as superior in one respect must be superior in all [Lovejoy 1936: 56f.], but that became the general impression later. Lovejoy says:

It will be seen that there was an essential opposition between two aspects of Aristotle's influence on subsequent thought, and especially upon the logical method not merely of science but of everyday reasoning. . . . He is oftenest regarded, I suppose, as the great representative of a logic which rests upon the assumption of the possibility of clear divisions and rigorous classification. Speaking of what he terms Aristotle's "doctrine of fixed genera and indivisible species," Mr. W. D. Ross has remarked that this was a conclusion to which he was led mainly by his "close absorption in observed facts." Not only in biological species but in geometrical forms . . . he had evidence of rigid classification in the nature of things. But this is only half of the story about Aristotle; and it is questionable whether it is the more important half. For it is equally true that he first suggested the limitations and dangers of classification, and the non-conformity of nature to those sharp divisions which are so indispensable for language and so convenient for our ordinary mental operations. . . .
From the Platonic principle of plenitude the principle of continuity could be directly deduced. If there is between two given natural species a theoretically possible intermediate type, that type must be realized—and so on *ad indefinitum*; otherwise there would be gaps in the universe, the creation would not be as "full" as it might be, and this would imply the inadmissible consequence that its Source or Author was not "good," in the sense which that adjective has in the *Timaeus*. [p. 57f.]

From Aristotle's notion of an ontological scale—the higher beings were less potential and more determinate (God, the *ens perfectissimum*, could not be otherwise than he is), and "all individual things may be graded according to the degree to which they are infected with [mere] potentiality" [Ross 1949: 178, after a discussion of the Metaphysics, quoted in Lovejoy on p. 59]. It is perhaps arguable if this idea really does exist in Aristotle's writings or that he intended it; but whether or not he did, this is the idea that was formulated by the neo-Platonists [Lovejoy 1936: 61–66, on Plotinus's construction of the chain], and various passages in his works led to a serial classification of organisms [Singer 1950: 40–41]. This philosophy influenced many biologists, especially, in this context, Bonnet, as we shall discuss in the next chapter. Later, in the late eighteenth century, at first in Buffon, then his pupil Lamarck and Erasmus Darwin in England, but also in Maupertuis, Robinet, Diderot, Holbach, and Kant, the Great Chain became temporalized. The ladder became a pathway.

The *lex completio*, and the *lex continui*, taken together, do not allow for temporal change. Leibniz could not allow that there could be more monads later than earlier [Lovejoy 1936: 256–258]. Others, most famously Voltaire, rejected the gradation scale for the same reason. Things did change, and they weren't always for the best. One or the other had to go. As we shall see with Lamarck, the *lex completio* was the first of the two major planks of the Great Chain to be rejected in biology. The Great Chain does imply that there is no real distinction between species, as all intermediate gaps are filled. This influenced Linnaeus, in his view of classification [Stevens 1994].

PETER RAMUS AND THE LOGIC OF WHOLES AND PARTS

Peter Ramus (1515–1572) was something like a protostructuralist, apparently not very consistent or even clear in his logical writings [Ong 1958].[2] However, he plays a role here for several reasons. One is that he mediates the anti-Aristotelian views of the Renaissance humanists like Valla to the post-Reformation age; and in so doing, he introduced to intellectual thought both a distinction between artificial and natural logic, but also the tree of Porphyry, which is occasionally referred to as the Tree of Ramus. A shameless self-promoter, he was censured by the religious authorities of Catholic France and was eventually killed in the St. Bartholomew's Day Massacre for having converted to Calvinism.

Ramus was greatly dependent on the terms in which medieval logic was discussed, but he had an idiosyncratic view of the genus-species distinction. He treated general terms as based on clusters or groups of simples, and he called these genera. He concluded that species were therefore the same thing as individuals or particulars, and he said that men and women were different species, for which he was condemned and in which he was followed by various late sixteenth-century authors. He based this apparently on the view that words had a single denotation, and that the denotation was of material accidents [p. 203]: "A species is the thing itself concerning which the genus answers [in replying to the question, What is it?]: thus the genus man answers concerning Plato, the genus dialectician concerning this particular dialectician" [*Training in Dialectic*, 1543, folio 14; quoted in Ong 1958: 204].

He seems to think that the general term *(genus)* is an answer to a given question of the constitution *(species)* of a particular thing. He has no notion of descending genera and species, nor that one genus (e.g., *dialectician*) might be a subaltern species in another *(man)*. Either he misunderstood the Scholastic tradition, which is Ong's interpretation, or he was redefining the logical enterprise. Ramus's ideas were influential for much of the following century and influenced the Port Royal Logic [Freedman 1993]. However, he seems to have had little effect on biological classification as it developed.

LEONHART FUCHS AND CONRAD GESNER: PICTURES, GENUS, AND SPECIES

Following the Renaissance, beginning in the sixteenth century after the decline in empirical work on animals and plants in the fourteenth century, modern biology begins to take root.[3] The period of the Reformation was largely lost to natural history, with the exceptions of the herbals of Leonhart Fuchs (1501–1566) and Otto Brunfels (1500–1534), and the bestiary of Conrad Gesner (1516–1665) and the naturalists who followed him such as Guillame Rondelet (1507–1556) and Pierre Belon (1517–1564), who is remembered for his study of the homology between bird and human skeletons. While Brunfels is regarded as derivative and Heironymus Fock as naive, Fuchs has some novelties that were later influential [Sachs 1890; Sprague and Nelmes 1928–1931]. He tended to use binomials (with the "genus" name often *after* the "species" name) and organized his kinds (*genera* and *species* are used indifferently) under heads *(capita)*. However, he has nothing resembling the Linnaean ranks, although his descriptions can be mapped to Linnaean species as Sprague and Nelmes [1928–1931] have shown. Moreover, he did not divide his "genera" evenly. For example, Sprague and Nelmes list his classification of *Ranunculus* (a flowering plant genus that includes the buttercups) with the corresponding Linnaean species (figure 3). Each of the species listed are listed in an apparently arbitrary order. There is no type species of the "genus."

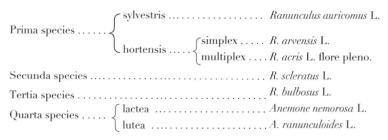

Figure 3 Fuchs on genera and species. An example of Fuchs's capita and general species [redrawn from Sprague and Nelmes 1928–1931: 557]. In other cases, he would list the primum genus and so on.

Of the bestiary writers of the Reformation, Gesner is most interesting, because like Frederick and Albert, he studied the organisms for himself. Of equal note is that he engaged woodcuts to be made of his specimens, allowing readers to identify which actual species he was talking about. His *Historia animalium* ran to 3,500 pages in four volumes, organized by Aristotelian categories of viviparous and oviparous organisms, then birds, fishes, and reptiles and insects [Nordenskiöld 1929: 93f.]. In his *Historia plantarum* (1548–1566), he summarized twenty years' work of direct descriptions and drawings of 1,500 plants, including dissections of flowers and fruit.

The formal distinction into genera and species in botany arises in the work of Gesner, according to Arber [1938: 166], when he employs the practice of giving genera substantive names, and Arber considers him the earliest to do so. However, his work was not widely known, and Arber instead considers that Fabius Columna (or Fabio Colonna, b. 1567–1650), in his *Ekphrasis* (1616), under the influence of Cesalpino, published the first views on the nature of genera in botany, relying on flower and seed rather than the older morphologies of leaf and stem to distinguish them.[4] Columba also used etching in botanical illustration in his 1592 herbal *Phytobasanos*, for the first time.

Although a Protestant and friend of Zwingli, Gesner was regarded well by all, except his rival Ulisse Androvani, who plagiarized him and sneered when referring to him [Raven 1953, chapter 5], although Androvani is regarded by Nordenskiöld as superior to Gesner in classification and methods, if not accuracy [Nordenskiöld 1929: 95]. Stresemann notes that while his arrangement of species was novel, diverging from the alphabetical order Gesner and the bestiaries and etymologies used, Aldovani's was no improvement, based on beak structure [Stresemann 1975: 22]. One of Gesner's friends was Caspar Bauhin, to whom we now turn.

ANDREAS CESALPINO AND CASPAR BAUHIN: THE BEGINNINGS

True investigation of nature consists not only in
deducing rules from exact and comparative observation
of the phenomena of nature, but in discovering the
genetic forces from which the causal connexion, cause
and effect may be derived. In pursuit of these objects, it
is compelled to be constantly correcting existing
conceptions and theories, producing new conceptions

and new theories, and thus adjusting our own ideas
more and more to the nature of things. The understanding does not prescribe to the objects, but the
objects to the understanding. The Aristotelian philosophy and its medieval form, scholasticism, proceeds in
exactly the contrary way; it is not properly concerned
with acquiring new conceptions and new theories by
means of investigation, for conceptions and theories
have been once and for all established; experience must
conform itself to the ready-made system of thought;
whatever does not so conform must be dialectically
twisted and explained till it apparently fits in with
the whole.
Julius Sachs [1890: 86]

Sachs's rather cynical view of medieval and renaissance botany does
an injustice to the writers themselves. For example, Luca Ghini
(1490–1556), who founded the botanical garden at Pisa for Cosimo de
Medici, introduced the practice of herbariums, in which plants were dried
and pressed so that exact identification of plants could be made. He
influenced a number of his contemporaries, including Gesner, Brunfels,
Fuchs, and Andreas Cesalpino, who succeeded him as director of the
garden. Although he did not write anything that has survived, he is
credibly the founder of empirical botany [Morton 1981: 121–124].

According to Lovejoy, the Florentine Andreas Cesalpino (1519–1603)
was an enthusiast of Aristotelian classification, and Lovejoy notes that it
was the fresh study of Aristotelian writings that set Cesalpino to producing
his *De Plantis* (1583).[5] In this and his *Peripatetic Problems* of 1588, under
the Latin title *Quaestionum peripateticarum, Libri V* [Nordenskiöld
1929: 113], he worked on strict Aristotelian lines: "exhaustive comparative analysis of the forms, concisely worded theoretical definitions, and,
based on these, abstract conclusions" [Nordenskiöld 1929: 192]. Sachs
[1890: 52] quotes him saying in chapter 13, "That according to the law
of nature like always produces like and that which is of the same species
with itself." In chapter 14, Cesalpino stated, "We seek similarities and
dissimilarities of form, in which the essence ('substantia') of plants consists, but not of things which are merely accidents of them ('quae accidunt
ipsis')" [quoted in Sachs 1890].

Cesalpino therefore seems to be a mediate source of both the idea
that species are fixed, and that species have an underlying essence, or

substance in the Latin, that is distinct from other characters that may vary accidentally. Whether this meant that he was a fixist, though, is open to doubt. Cesalpino, like all those of the time, was concerned to differentiate *genera et species* in *De plantis*, by identifying the botanical characters *(differentiae)* and he uses the fructification of the plants as his basis [Morton 1981: 135–140]. He says in the preface to *De plantis* [translated in Morton 1981: 135], "Since science consists in grouping together of like and the distinction of unlike things, and since this amounts to the division into genera and species, that is, into classes based on characters *(differentiae)* which describe the fundamental nature of the things classified, I have tried to do this in my general history of plants." He specifically names Theophrastus as the authority for this methodology.

Independently of Cesalpino, Caspar (or Kaspar) Bauhin of Switzerland (1550–1624) organized known scientific names and descriptions of plants into two works *(Prodromus* and *Pinax theatri botanici)*, in which the genus-species arrangement was used. However, he did not necessarily use it consistently as Linnaeus did [although Sachs thinks he was quite consistent, 1890: 33–35, and is dismissive of Linnaeus's "dry systematising manner," p. 40]. Even so, Bauhin used the common likeness of plant forms as his basis for classification, unlike Cesalpino's artificial system, which was based on the special characters of what he considered the "soul" or heart of the plants (hearkening to Aristotle's *De Anima*). In his *Phytopinax* (1596), Bauhin states in the preface that he has applied one name to each plant for clarity [Arber 1938: 168]. Cesalpino tended, according to Nördenskiold [p. 193f.], to focus on the fruits of plants in his classifications, a decision that also later influenced Linnaeus. Both Cesalpino and Bauhin were transitional between the older herbals tradition in which classification was either by alphabet or by utility in medicine and cooking.

THE UNIVERSAL LANGUAGE PROJECT

During the Renaissance, philosophers were more concerned with souls (especially the rational soul) and knowledge than species, but toward the end neo-Platonism was revived, particularly at Cambridge University under the general leadership of Ralph Cudworth and the later title of the "Cambridge Platonists." This movement was instrumental in the establishment of seventeenth-century science in England in particular, and Bishop John Wilkins, among others, began discussion groups to review recent experiments and results. One of the movements they continued was that of the search for a logical system that could ensure a perfectly expressive and complete language [Slaughter 1982; Rossi 2000].

Mary Slaughter [1982] has discussed the Universal Language Project at length, and she traces the influence of this approach from the medieval herbalists and the heritage of Aristotle. She noted that the sixteenth and seventeenth centuries inherited an "essentially Aristotelian" worldview that determined the way people thought about language [p. 87]. The theory of language that they developed is founded on the epistemology of Aristotelian scholasticism—based on analytic differentiation of the *topoi*, the Categories. The "universal grammarians" Vossius, Caramuel, and Campanella of the sixteenth century in turn gave way to the "universal language projectors" of the seventeenth. These began with the work of Francis Bacon's, *The Advancement of Learning* (1605), in which, Slaughter says, he systematically "described, analyzed and classified Renaissance provinces of knowledge, dividing learning into the arts of imagination, memory and reason . . . . Among the subdivisions of the arts of reason was rhetoric; among its parts Bacon included the Art of Transmission, or communication" [p. 89]. Bacon argued that letters of the Latin alphabet are conventional signs, but the Chinese characters were "real characters, not nominal," which he said "represent neither letters nor words, but things and notions; insomuch that a number of nations whose languages are altogether different, but who agree in the use of such characters . . . communicate with each other in writing; to an extent that indeed any book written in characters of this kind can be read off by each nation in their own language" [quoted on p. 90]. Words are signs that can represent the world, or they can be false and misleading, which he called "Idols of the Marketplace," "idols which have crept into the understanding through the alliances of words and names" [*Novum Organum*, 1620, I.lix, quoted on p. 91]. Slaughter comments that for Bacon linguistic problems merely reflect conceptual problems [p. 92]. He wrote that "[t]here is no soundness in our notions whether logical or physical. Substance, Quality, Passion, Essence itself, are not sound notions; much less are Heavy, Light, Dense, Rare, Moist, Dry, Generation, Corruption, Attraction, Repulsion, Element, Matter, Form and the like; but all are fantastical and ill defined" (*Novum Organum* I.xv). Bacon goes on to say that we can be fairly sure that words for animal species and simple sensory perceptions are accurate enough:

> Our notions of less general species, as Man, Dog, Dove and of the immediate perceptions of the sense, as Hot, Cold, Black, White, do not

materially mislead us; yet even these are sometimes confused by the flux and alteration of matter and mixing of one thing with another. All the others which men have adopted are but wanderings, not being abstracted and formed by proper methods. [*Novum Organum* I.xvi]

So, there is a method *(organon)* one must adopt to do this properly, says Bacon:

There are and can be only two ways of searching into and discovering truth. The one flies from the senses and particulars to the most general axioms, and from these principles, the truth of which it takes for settled and immovable, proceeds to judgment and to the discovery of middle axioms. And this way is now in fashion. The other derives axioms from the senses and particulars, rising by a gradual and unbroken ascent, so it arrives at the most general axioms last of all. This is the true way, but as yet untried. [*Novum Organum* I.xix]

Bacon, in effect, treats science as a bottom-up process of generalization, in opposition to the top-down classification practiced by the Aristotelians. As Slaughter comments, enumeration under Bacon's methodology is followed by classification, which is the "critical operation in the discovery and the definition of the essences of things. Natures could be properly defined only when instances were properly classified into 'the true divisions' of nature" [p. 93]. For him and his contemporaries, real science meant that the labels of things were correct representations or signs of the way the world was and that essences were found as axioms, or generalizations, made step-by-step from empirical inductions.

Bacon recognized that organisms and other things could deviate from the type. There is no necessity for things to be constrained by their species' essences:

Among Prerogative Instances I will put in eighth place *Deviating Instances*, that is, errors, vagaries, and prodigies of nature, wherein nature deviates and turns aside from her ordinary course. Errors of nature differ from singular instances in this, that the latter are prodigies of species, the former of individuals. Their use is pretty much the same, for they correct the erroneous impressions suggested to the understanding by ordinary phenomena, and reveal common forms. . . .

[W]e have to make a collection or particular natural history of all prodigies and monstrous births of nature; of everything in short that is in nature new, rare, and unusual. [II.xxix]

Bacon also thinks it worthwhile to discover which things are a mixture of species [II.xxx]; but it is clear that he considers the term in a much broader manner than the modern sense, and his notion of hybrids in the final phrase is quite arbitrary:

> I will put in the ninth place *Bordering Instances*, which I will also call *Participles*. They are those which exhibit species of bodies which seem to be composed of two species, or to be rudiments between one species and another. . . .
>
> Examples of these are: moss, which holds a place between putrescence and a plant; some comets, between stars and fiery meteors; flying fish, between birds and fish; bats, between birds and quadrupeds; also the ape, between man and beast—
>
> Simia quam similes turpissima bestia nobis;[6]
>
> Likewise the biformed births of animals, mixed of different species, and the like. [Bacon 1960: 178f.]

Elsewhere, in the *New Atlantis* (1627), he has his mythical sage enumerate the things his utopian society can do with animal breeding in Salomon's House: "we make them differ in colour, shape, activity, many ways. We find means to make commixtures and copulations of divers kinds, which have produced many new kinds, and them not barren, as the general opinion is." Although he does not think nonsterile hybridization is impossible in forming new species, species are only mutable in a limited way.

He was followed by Descartes, who proposed that an artificial language could be constructed on the true divisions of nature (in his letters), but the work of the linguist Comenius in *Janua linguarum researata* (1631) and *Janua linguarum researata vestibulum* (1633) made a start at a universal language answering to all things. In later work, he expanded the scope of this project, and he had unrealized plans for "an inductive history in which all things all things, which have ever been exactly observed and proved beyond all possibility of mistake to be true, are faithfully collected and set before our eyes; so that by an adequate examination of each one and by comparison of one with another the Universal Laws themselves of Nature may be brought within our knowledge" [quoted in Slaughter 1982, p. 102f.]. Descartes tried, unsuccessfully, to engage the botanist and polymath Joachim Jung to edit his works. Other universal language projectors included Samuel Hartlib, Theodore Haak, Francis Lodowyck, Cave Beck, Francis Van Helmont, George Dalgarno, Athanasius Kircher, Johann Joachim Becher, Seth Ward, and various others involved in the nascent Royal Society at Cambridge [cf. Rossi 2000].

One contributor, one might say the zenith, of the Universal Language Project was Bishop John Wilkins (1614–1672). My namesake was the brother-in-law of Oliver Cromwell and is widely regarded as the founder of the Royal Society [Slaughter 1982; Wilkins 1970; Wright Henderson 1910], of which he was the first secretary. He produced *An Essay towards a Real Character and a Philosophical Language* (1668), the only long-standing influence of which is said to have been the foundation of the scheme for Roget's *Thesaurus* (and the occasion for a Borges essay). In fact, Wilkins's scheme was not so much a universal language as a kind of indexical classification of concepts of his time (which, in typical style for the time and since, he thought to be universally true of the entire human condition), where words were actually unique keys to each conceptual "address." He managed to engage (for money) a naturalist by the name of John Ray to produce a table of species according to his own a priori categories. Ray felt that this a priori scheme was too constrictive; but, according to Raven [1986: 182f., 192], it did provide him with a motivation to do better in his later publications (see below). Ray also translated the *Essay* into Latin and remained a close friend to Wilkins until his untimely death from kidney stones, despite his objections to the underlying philosophy of the project. The Latin edition was never published [Clauss 1982].

Out of the neo-Platonic resources of the Cambridge Platonists and the Universal Language Project, the transition was begun to an autonomous program of biological classification. This heritage had two opposing aspects. First, as we have seen, there was the history of species as the sharp categories of a top-down classification. The other, via Descartes and Leibniz, was the Great Chain of Being notion of continuous gradation from simpler forms to more complex. Ironically, both of these develop ideas nascent or explicit in Aristotle's writings. Lovejoy [1936: 227] writes:

> The first [aspect] made for sharp divisions, clear-cut differentiations, among natural objects, especially among living beings. To range animals and plants in well-defined species, presumably (since the Platonic dualism of realms of being was still influential) corresponding to the distinctness of Eternal Ideas, was the first business of the student of the organic world, The other tended to make the whole notion of species appear a convenient but artificial setting-up of divisions having no counterpart in nature. It was, on the whole, the former tendency that prevailed in early modern biology.

In the tension between sharp classification and gradual variation from one form to another in the Great Chain, we see the early stages of natural species realism, based on typological definitions, and species nominalism, based on the unreality of any divisions between them. Much of the early biological debate over species is an attempt to deal with this tension, and indeed it continues to the present day. This is made more complicated by the second main developing tradition of the distinction between "natural" and "artificial" classification.

JOHN LOCKE AND GOTTFRIED LEIBNIZ: REAL AND NOMINAL ESSENCE

For the natural tendency of the mind being towards knowledge; and finding that, if it should proceed by and dwell upon only particular things, its progress would be very slow, and its work endless; therefore, to shorten its way to knowledge, and make each perception more comprehensive, the first thing it does, as the foundation of the easier enlarging its knowledge, either by contemplation of the things themselves that it would know, or conference with others about them, is to bind them into bundles, and rank them so into sorts, that what knowledge it gets of any of them it may thereby with assurance extend to all of that sort; and so advance by larger steps in that which is its great business, knowledge. This, as I have elsewhere shown, is the reason why we collect things under comprehensive ideas, with names annexed to them, into genera and species; i.e. into kinds and sorts.

**John Locke, *Essay on Human Understanding*
[Book II, chapter 32, §6]**

A friend of the Cambridge Platonists and universal language projectors was John Locke. But Locke was also a friend of Robert Boyle, the atomistic ("corpuscularean") chemist, whose philosophy demoted sensory perceptions from immediate empirical experiences of things to secondary qualities, since we do not observe the corpuscles of which things are composed. Locke followed Boyle rather than Bacon in this, and so our knowledge of things must be nominal rather than essential [Slaughter 1982: 193–207]. Locke is sometimes called a nominalist, and there is a sense in which this is true, but it relates to his views about names, and not about the underlying things the names refer to. Names denote abstractions, and some names denote the essences of ideas. But Locke does not deny that there are what he calls "real essences," only that the essences of kinds, or "sortals," as he calls them, agree to anything else but "nominal essences."

Locke is not particularly remembered by biologists for his contribution to the species debate, but he should be [Cain has taken some steps in this direction, 1997]. His views, expressed especially in chapters 3 to 7 of Book III of the *Essay*, are the first statement of a position one may call *species conventionalism*, which is held today even by those who reject his essentialistic notion of names. He triggered a response by Leibniz and was influential on French naturalists, including Buffon and Lamarck:

> The learning and disputes of the schools having been much busied about genus and species, the word essence has almost lost its primary signification: and, instead of the real constitution of things, has been almost wholly applied to the artificial constitution of genus and species. It is true, there is ordinarily supposed a real constitution of the sorts of things; and it is past doubt there must be some real constitution, on which any collection of simple ideas co-existing must depend. But, it being evident that things are ranked under names into sorts or species, only as they agree to certain abstract ideas, to which we have annexed those names, the essence of each genus, or sort, comes to be nothing but that abstract idea which the general, or sortal (if I may have leave so to call it from sort, as I do general from genus), name stands for. And this we shall find to be that which the word essence imports in its most familiar use. [Book III, chapter 3, §15]

Locke considers "species" to be merely the Latinized version of the good English word *sort* or *kind* [Book III, chapter 1, §6] and held that species are conventional names used mainly for specialists to communicate. In fact, most of the then-current species names were based on what we would now call "folk taxonomy":

> *This shows Species to be made for Communication.*—The reason why I take so particular notice of this is, that we may not be mistaken about *genera* and *species*, and their *essences*, as if they were things regularly and constantly made by nature, and had a real existence in things; when they appear, upon a more wary survey, to be nothing else but an artifice of the understanding, for the easier signifying such collections of *ideas* as it should often have occasion to communicate by one general term; under which divers particulars, as far forth as they agreed to that abstract *idea*, might be comprehended. And if the doubtful signification of the word *species* may make it sound harsh to some, that I say the species of mixed modes are "made by the understanding"; yet, I think, it can by nobody be

denied that it is the mind makes those abstract complex *ideas* to which specific names are given. And if it be true, as it is, that the mind makes the patterns for sorting and naming of things, I leave it to be considered who makes the boundaries of the sort or *species*; since with me *species* and *sort* have no other difference than that of a Latin and English *idiom*. [Book III, chapter 5, §9]

But supposing that the *real essences* of substances were discoverable by those that would severely apply themselves to that inquiry, yet we could not reasonably think that the *ranking of things under general names was regulated by* those internal real constitutions, or anything else but *their obvious appearances*; since languages, in all countries, have been established long before sciences. So that they have not been philosophers or logicians, or such who have troubled themselves about *forms* and *essences*, that have made the general names that are in use amongst the several nations of men: but those more or less comprehensive terms have, for the most part, in all languages, received their birth and signification from ignorant and illiterate people, who sorted and denominated things by those sensible qualities they found in them; thereby to signify them, when absent, to others, whether they had an occasion to mention a sort or a particular thing. [Book III, chapter 6, §25]

Locke here recognizes a distinction between folk taxonomy and a proper (philosophical) enquiry into the scientific issues. Leibniz in the *New Essay on Human Understanding* [1996, p. 319] paraphrases this more succinctly (Philolethes is Locke; Theophilus is Leibniz):

> *Philolethes*: §25. Languages were established before sciences, and things were put into species by ignorant and illiterate people.

To which *he* responds:

> *Theophilus*: This is true, but the people who study a subject-matter correct popular notions. Assayers have found precise methods for identifying and separating metals, botanists have marvelously extended our knowledge of plants, and experiments have been made on insects that have given us new routes into the knowledge of animals. However, we are still far short of halfway along our journey.

Leibniz is more of an optimist about the causal powers that form species being available to investigators than Locke, who seems to propose a permanent conventionalism based on the current inability to define species according to their "internal real constitutions." Nevertheless, Leibniz held to a view that species were not real as discrete objects. He was an adherent of the *scala naturae*, or the Great Chain of Being, and popularized the *lex completio*—the view that there could be no incompleteness in the world as made by a beneficent God.

Locke discussed, rather interestingly, the presence in species of divergences from the type [Book II, chapter 6, §§16–17, 26–27]. In this he was preceded by Cusa, as we have seen, and also by Francis Bacon in the *New Organon*, Book II, section 29 [New Organon, Book II, §29; cf. Glass 1959a: 36], but Locke's discussion is surprisingly modern in a way that Bacon's is recognizably medieval. He greatly influenced, among others, Buffon, through the writings of French admirers of Locke's empiricism and new way of ideas (here again, the appearance of *eidos*). He is often thought to be solely a species nominalist, as Buffon transitorily was, but it seems more accurate to say that he believed our ideas and associated names were conventional, but that, as Bacon thought, there was some underlying essence that was likely to remain out of our reach. This idea was famously given its canonical expression in the doctrine of the *noumenal* and the *phenomenal* of Kant.

Locke's views on relations [*Essay*, Book II, chapter 28, §2] are also significant—although he rejects the idea that relations are outside the understanding, he nevertheless has a category of "natural relations" that apply to genealogical relationships between organisms. Although we only have access to the ideas of these simple and mixed modes, they are nevertheless something in the natural world. He does assert here that the *words* used for the relations, such as *father*, *brother*, and *cousins-germaine*, are conventional, though, even if the relations are not. Relations resurfaced in the writings of William Hamilton [Hamilton, Mansel, and Veitch 1874, vol. 2], Russell and Whitehead [Whitehead and Russell 1910], and throughout the middle of the twentieth century.

JOHN RAY: PROPAGATION FROM SEED

John Ray (1627–1705) was a seventeenth-century naturalist who, in conjunction with nobleman Francis Willughby (1635–1672), prepared the first systematic flora for a region—at first of Cambridgeshire, and later of Britain [Raven 1986]. In the *Historia plantarum* (1686–1704), he and Willughby attempted to describe all known species of plants. He also collaborated with Willughby, before the latter's untimely death, on a treatment of insects, animals, and fishes. To this end, Ray needed to define *species*, and he was the first to do so entirely in a biological context. This strongly influenced Linnaeus's conceptions of species and other ranks.

In the *Historia plantarum generalis*, in the volume published in 1686, Ray defined a species as follows:

> In order that an inventory of plants may be begun and a classification of them correctly established, we must try to discover criteria of some sort for distinguishing what are called "species." After long and considerable investigation, no surer criterion for determining species has occurred to me than the distinguishing features that perpetuate themselves in propagation from seed. Thus, no matter what variations occur in the individuals or the species, if they spring from the seed of one and the same plant, they are accidental variations and not such as to distinguish a species. . . . Animals likewise that differ specifically preserve their distinct species permanently; one species never springs from the seed of another nor vice versa. [quoted in Mayr 1982: 256]

Mayr notes, "Here was a splendid compromise between the practical experience of the naturalist, who can observe in nature what belongs to a species, and the essentialist definition, which demands an underlying shared essence" [Mayr 1982: 257]. However, it seems to me the influence of the generative conception is more apparent here. Ray is dealing with the Aristotelian problem of accidental variation—like Locke, he believes there must be a real essence, and not merely a nominal one. He conjectures that it is, or rather defines it to be, based on descent by generation. But in most cases, as many have observed since (see Buffon, discussed later), descent is *not* observable; and he clearly did not observe the fact that one species never springs from the seed of another. In fact, he wrote, in the *Methodus plantarum* of 1682:

> I would not have my readers expect something perfect or complete; something which would divide all plants so exactly as to include in positions anomalous or peculiar; something which would so define each genus by its own characteristics that no species be left, so to speak, homeless or be found common to many genera. Nature does not permit anything of the sort. Nature, as the saying goes, makes no jumps and passes from extreme to extreme only through a mean. She always produces species intermediate between higher and lower types, species of doubtful classification linking one type with another and having something common with both—as for example the so-called zoophytes between plants and animals. [quoted in Glass 1959a: 35][7]

According to Glass, Ray did allow some limited transmutation between related species, especially hybridization, a problem that Linnaeus later also had to accommodate. Sometimes called the originator of the British natural theology tradition[8]—he wrote the book *Wisdom of God Manifested in the Works of the Creation* [1691; Gould 1993: 140]—

Ray also had a problem with extinction. It was not that he did not find it an operational concept; it was ruled out in terms of the principle of plenitude [Bowler 1989a]. Initially he thought that fossil forms would be found alive elsewhere, but later he was forced into denying that fossils were even the remnants of living forms but had instead grown within the rocks.

Ray is responsible for formulating the first explicit entirely *biological* notion of species, but this is not the same as saying that he presented what we would now consider a biological species concept. For him, this was a sense of "species" that applied to reproducing *forms*—that is, living things. It was the first time a concept was proposed that applied *only* to the classification of living things. Prior to this, all such concepts were general-duty concepts of classification that were then applied equally to, say, books or rocks as to life.

Ray clearly saw his classificatory logic as the continuation of the Aristotelian, Scholastic, tradition, but his adherence to the Great Chain of Being, its neo-Platonist heritage, and the idea that there needed to be a real essence led him to propose not so much an operational concept, as Mayr would have it, but a metaphysical one. As mentioned, Ray had prepared a table of species before the *Historia* for Bishop John Wilkins's magnum opus *Essay toward a Real Character and Philosophical Language*, published by the Royal Society in 1668. As we have seen, Wilkins was one of the so-called Cambridge Platonists, a neo-Platonic school that was directly and indirectly responsible for inspiring much scientific work done in Britain at the time and from which the Royal Society sprang. Like Wilkins, Ray was influenced by Ralph Cudworth, whose ideas were consciously in the school of Plotinus and Porphyry. In addition, Ray had received the standard *Trivium* education (grammar, logic, and rhetoric) that was presented at the grammar schools of the time, which at sixteen and a half he began also to receive from Cambridge. With his friend and mentor Henry More, Ray was thoroughly inculcated in the techniques and terminology of the older Scholastic logic. Although he accepted the *scala naturae*, he nevertheless held to the fixity of species since creation, a view that he bequeathed to Linnaeus. That Ray held to fixity of species as created by God is illustrated by this comment, made in a letter: "[T]he number of species being in nature certain and determinate, as is generally acknowledged by philosophers, and might be proved also by divine authority, God having finished his works of creation, that is, consummated the number of species in six days" [quoted in Greene 1959: 131].

NEHEMIAH GREW: SPECIES AS ESSENTIAL KINDS

Almost contemporaneous with Ray was Nehemiah Grew (1641–1712), who undertook to bring the anatomical studies of plants under the scope of microscopic studies. Grew's notion of species was fairly standard, but what is remarkable, apart from the exquisite figures of cell structure, in his treatment is his justification of the use of classification as a way of making inductive generalizations. His *The Anatomy of Plants* [Grew 1682] is a series of lectures given before the Royal Society, including his general overview of botany ("vegetables"—Grew is primarily interested in crop plants), *An Idea of a Philosophical History of Plants*, given on January 8 and January 15, 1672:

> 12. § For in looking upon divers *Plants*, though of different *Names* and *Kinds*; yet if some affinity may be found betwixt them, then the *Nature* of any one of them being well known, we have thence ground of conjecture, as to the *Nature* of all the rest. So that as every *Plant* may have somewhat of *Nature individual* to it self; so, as far as it obtaineth any *Visible Communities* with other *Plants*, so far, may it partake of *Common Nature* with those also. [p. 6]

In short, Grew is arguing that if we can identify the nature of one kind, we can expect that others will share in it to the extent that they share an "affinity" with it. Nature is here a kind of inductive warrant, because genera share their own properties, and kinds will therefore also share the generic properties. He expands on this later, noting that if we know the essence of a plant (the causes of its being and becoming what it is), we can know what is necessary to it and what is accidental—based on its internal structure as well as its outward appearance:

> 20. § From all which, we may come to know, what the *Communities* of *Vegetables* are, as belonging to all; what their *Distinctions*, to such a Kind; their *Properties*, to such a Species; and their *Peculiarities*, to such Particular ones. And as in *Metaphysical*, or other Contemplative Matters, when we have a distinct knowledge of the *Communities* and *Differences* of Things, we may then be able to give their true *Definitions*: so we may possibly, here attain, to do likewise: not only to know, That every *Plant* Inwardly differs from another, but also wherein; so as not surely to Define by Outward *Figure* than by the Inward *Structure*. What that is, or those things are, whereby any *Plant*, or Sort of *Plants*, may be distinguished from all others. And having obtained a knowledge of the *Communities* and *Differences* amongst the *Parts* of *Vegetables*; it may conduct us through a *Series* of more facile and probable *Conclusions*, of the ways of their *Causality*, as to the *Communities* and *Differences* of *Vegetation*. [p. 10]

So essence for Grew is a matter of causally efficacious internal microstructure, which we would now regard as a natural kind. This makes him one of the few actual material essentialists we shall encounter in the history of biology before the nineteenth century. He employs the Aristotelian definitional approach, naturally, but treats the definitions as identifying the true causes, the "active capacitating Causes," of the nature of the plants.

53. § The prosecution of what is here proposed, will be requisite, To a fuller and clearer view, of the *Modes* of *Vegetation*, of the *Sensible Natures* of *Vegetables*, and of their more Recluse *Faculties* and *Powers*. First, of the *Modes* of *Vegetation*. For suppose we were speaking of a *Root*; from a due consideration of the *Properties* of any *Organical Part* or *Parts* thereof; 'tis true, that the real and genuine *Causes* may be rendred, of divers and other dependent *Properties*, as spoken generally of the whole *Root*. But it will be asked again, What may be the *Causes* of those *first* and Independent ones? Which, if we will seek, we must do by inquiring also, What are the *Principles* of those *Organical Parts*? For it is necessary, that the *Principles* whereof a Body doth consist, should be, if not all of them the *active*, yet the *capacitating Causes*, or such as are called *Causae sine quibus non*, of its becoming and being, in all respects, both as to *Substance* and *Accidents*, what it is: otherwise, their Existence, in that Body, were altogether superfluous; since it might have been without them: which if so, it might then have been made of any other; there being no necessity of putting any difference, if neither those, whereof it is made, are thought necessary to its Being. [p. 20]

Rather clearly, Grew thinks of these causes as developmental and hence generational. The outer—or morphological, as we would now say— appearances may lead us to uncover those developmental properties. Grew thought these were to be found in the microscopic anatomy and physiology of the organisms, and his illustrations of the microstructures of the leaves, stalk, and so on, remain exceptionally accurate.

JOSEPH PITTON, DE TOURNEFORT: NAMES FOR DIFFERENCES

In his *Botanical Institutions* (*Institutiones rei herbariae*, 1700, earlier published in 1694 as *Elemens botanique*), Joseph Pitton, de Tournefort [1628–1708, published in English as Tournefort 1716–1730: 2] held that it did not really matter whether one diagnosed a species or a variety: "I not only enumerate the several Species of Plants, but often mention what the Botanists call Varieties; not at all solicitous whether they be really the same Species only varied and somewhat diversified, for as they differ in some sensible Qualities, they ought to be distinguished by peculiar Titles." He used both flowers and fruits to define species, in contrast to the Linnaean artificial account based on just the sexual apparatus of the plant. Tournefort did not define species, but he did use many different characters—the root, leaves, stalk, branch—to identify the differentia between species of a genus. However, he also used, but did not rely on, accidents such as color, taste, scent, size, site, and similarity to popularly known objects (p. 64 in the Latin edition). Tournefort clearly thought only of species as definitional and nomenclatural items.

CARL LINNAEUS: SPECIES AS THE CREATOR MADE THEM

Like Ray, the traditional logic was also taught to a poor Swedish student named Carl Linnaeus.[9] Born in 1707, Linnaeus died in 1770, the most celebrated Swede of his day. In 1761, he was knighted with the vernacular name Carl von Linné and took the Latinized name Carolus Linnaeus.

Linnaeus was a botanist and trained in Holland, where he published his first botanical works. Before Linnaeus, species were given all kinds of descriptive names, usually in Latin, up to about ten words long. Each author made up their own terms, and there was no real convention for referring to species. On Linnaeus's account, both species and genera were fixed, real, and known by definitions. He apparently believed that the genus was more real than the species, and he allowed late in life that species may occasionally arise, but only within genera, through hybridization. Some [e.g., Stafleu 1971; Mayr 1969, 1982] consider Linnaeus to be an essentialist regarding species. This was due to the fact that, unlike the medieval logical conception, for Linnaeus all species (at least in botany, zoology and mineralogy) were infimae species. He attempted to provide a diagnostic definition for each species, although his practice and adopted motto "*In scientia naturali principia veritatis observationibus confirmari debent*" [in natural science, the principles of truth ought to be confirmed by observation; Stafleu 1971] suggests that he was not firmly wedded to a priorism. In the 1735 *Systema Naturae* [10th ed., 1759, p. 7; Linne 1956], Linnaeus proposed a system of five ranks under the *summum genus* of the Empire of Nature *(Imperium Naturae)*:

> Regnum (Kingdom),
> Classis (Class),
> Ordo (Order),
> Genus,
> Species,

the last three of which corresponded to the logical ranks of *genus intermedium*, *genus proximam*, and *(infimae) species*. He also added a subspecific category of *Varietas*, which was the logical *individuum*. Later taxonomic conventions added the ranks of Phylum between Kingdom and Class, and Family between Order and Genus, giving seven ranks.[10] The philosophical notion of species was not entirely helpful in botany, so Linnaeus changed it a little. Instead of any number of subaltern genera, he made the scale of classes absolute, and instead of working downward, he started in the middle (at the genus). Linnaeus's ranks began at species, and these existed in genera. Hence, to name a species you needed to give the generic name and the species name. Humans are members of the genus *Homo* (or Man; according to Linnaeus, one of several[11]), and our species is called *sapiens* (the wise one). So in Latin our "name" is "the wise man." Humans, under his initial system, are

> Animals *(Regnum Animale)*,
> Mammals *(Classis Mammalia)*,
> Primates *(Ordo Primates)*,
> Man *(Genus Homo)*,
> Wise or rational *(Species sapiens)*.[12]

The "rational animal" definition of medieval logical taxonomies is evident [Broberg 1983]. What Linneaus did differently was to make species and genera fixed ranks. He established this universal system for the naming and classification of all organisms. There were, for example, various kingdoms—plants (Plantae) and animals (Animalia). Each species had a street address (its generic name, or *genus*) and a street number (its species name, or *epithet*).[13] Now, taxonomists (those who classify taxa, or groups of organisms[14]) could use a single and relatively simple system for their organisms, and all could agree on how to name them and what to name.

Linnaeus was definitely a special creationist—that is, he believed that each species was created specially by God—and Haller famously said of him that he thought himself a "second Adam" [Ramsbottom 1938: 195n.]. His project of classification relied heavily on the logic of Scholastic philosophy, which he employed in ways reminiscent of the Universal Language Project, to impose order on the world. He said once to his friend Sauvages that he was unable "to understand anything that is not systematically ordered" [Lindroth 1983: 23], and he was obsessive in this regard. To this end, he established a system that did impose that order on the living and nonliving world.

Linnaeus's widely cited "definition" of *species*, repeated throughout his writings in various wordings, is "There are as many species as the Infinite Being produced diverse forms in the beginning" (Species tot sunt diversae quot diversas formas ab initio creavit infinitum Ens; Fundamenta botanica No. 157, 1736). This is not really a definition but more of a statement of piety. However, in 1744 he was forced to allow that some species are the result of hybridization, at least in plants, because he observed it happening. A species of plant he placed in a genus *Peloria* (from the Greek *pelor*, meaning "monstrosity") was in stem and leaf structure part of the *Linaria* genus, but the flower was clearly different [Hagberg 1952: 196f.; Glass 1959b]. This admission was widely known by subsequent writers [e.g., Lee 1810; Gray 1821]. Still, he thought that genera were real and the possibilities for change limited. According to Larson [1967], Linnaeus imagined in the *Fundamenta fructifications* "that God created one species for each natural order of plants differing in habit and fructification from all others. These species, mutually fertile, gave birth to as many genera as there were different parents, their fructification somewhat changed" [p. 317].

In the *Pralectiones* (Lectures; 1744), Linnaeus went further:

> The principle being accepted that all species of one genus have arisen from one mother through different fathers, it must be assumed:
>
> 1. That in the beginning the Creator created each natural order only with one plant with reproductive power.
> 2. That by their various mixings different plants have arisen which belong to the mother's natural order as they are similar to the mother with regard to their fructifications, and are, as it were, species of the order, i.e., genera.
> 3. We may assume that plants have arisen within the orders, i.e. by genera of one order, may mix with each other. In this way there will arise species that should be referred to the mother's genus as her daughters.
> [quoted in Larson 1967]

Linnaeus thus employed the Great Chain of Being in a rather unusual way. Most "chainists" accepted what was later called the Principle of Plenitude (the *lex completio*), which stated that God would create everything that could be created, since he would not make an incomplete creation [Lovejoy 1936; Glass 1959b]. This usually meant that species graded into each other in a series of varieties. Linnaeus instead represented species using the metaphor of countries adjoining each other (in the *Philosophia botanica*, §77). In his early writing, all the territory is regarded as pretty much filled—as he said, nature does not make jumps—but the countries are discrete and distinct from one another. In the later work, this strict fixism of the first edition of the *Systema Naturae* has been modified. All hybrids did was fill in a rare empty bit of territory in God's time and plan. The borders were set by the genera, and all genera arose from a single species created by God. At the end of the 1750s, says Hagberg [1952: 199], Linnaeus was in a state of perplexity with respect to species. In 1755, he published *Metamorphosis plantarum*, dealing primarily with the development of plants but also with monstrosities and varieties. Such later hybrids he called the "children of time" in an anonymous entry in a competition at St. Petersburg in 1759 [Hagberg 1952: 201f.] and also in the *Species plantarum* (1753; 2nd ed., 1762–1763), where he speculated that a species of *Achillea* (yarrow, or staunchweed) *alpina* might have formed from another, *ptarmica*: "*An locus potuerat ex praecedenti formasse hanc?*" ["Could this have been formed from the preceding one by the environment?" in Volume II, 1266, of the 2nd ed., quoted in Greene 1959: 134]. Hagberg says, "Linnaeus never succeeded in pin-pointing his new conception of species. But the old one, that formed the basis of *Systema Naturae*, was utterly and irrevocably abandoned." In fact, reports Lindroth [1983], he started seeing hybrids everywhere, even where they were not.

Moreover, Linnaeus also noted that species grew differently according to the conditions of their locale. Of the genera *Salix*, *Rosa*, *Rubus*, and *Hieracium* (willows, roses, brambles, and hawkweeds), Linnaeus said that their description was problematic because of variability ("metamorphosis") of form in different soils and climates [Ramsbottom 1938: 200f.]. Habitat-induced variability will become an issue under Göte Turesson's investigation in the early twentieth century (discussed later). Linnaeus also experimented on propagating a hybrid geranium, with success, in 1759 [Ramsbottom 1938: 210f.]; he believed that maternal influences of hybrids affected the "medullary substance" and fructification of plants, but the leaf structure was due to the paternal species As time went on, he removed the statement that there were no new species from his 1766 edition of the *Systema Naturae* and crossed out the statement *natura non facit saltum* from his own copy of his *Philosophia Botanica*. A full list of Linnaeus's various pronouncements on species can be found in Ramsbottom [1938].

When Linnaeus was working, European trade and exploration were limited. Linnaeus himself classified around six thousand species of mainly Mediterranean and northern European plants, and later animals [Stafleu 1971]. This was more than had been done before, but still it was a fraction of what we know today. His students and adherents sent him specimens from around the world, and there was a steady "trade" in specimens between him and other taxonomists and collectors [Müller-Wille 2003]. Linnaeus hoped that his system would enable taxonomists to list all actual species, but he also knew that his system was artificial—that is, not the pure result of studying the actual characters of organisms, but also imposing an a priori scheme on them for convenience. He hoped there *would* be a "natural" scheme developed on the basis of an aggregation of characters, but he was never able to do more than a partial sketch of one. In his later work, he set up a "rational" system that allowed for there to be 3,600 genera in plants, each of which could generate species through hybridization. Although this was supposed to be a "natural" system (one based on the closeness of resemblance of all traits of the organisms and not just a single character), in fact, he chose just three features of plants and restricted the varieties to sixty types of each (hence $60^3 = 216,000$ maximum of plant species). However, this was fragmentary and in an appendix, and not developed further.

In summary, Linnaeus proposed a five-rank taxonomic system, and there were only a set number of species possible, although later he was forced by various observations, including his own, to accept that new species could be created through hybridization. All that remains of his taxonomic enterprise now are the names and general ranks of his system, but even this has been dramatically modified, with such groups as tribes, subfamilies, and so on, being added to deal with the massive increase in species discovered since. There are now as many as eighteen ranks or more, each with supra- and subranks, in the taxonomic hierarchy that is called "Linnaean" [cf. Mayr and Ashlock 1991].

Linnaeus distinguished between the diagnostic characters *(characters)* and actual traits *(notae)* of organisms, but it seems not much came of this distinction. He appears to have despaired of a natural system in his foreseeable future, and so he promoted a purely diagnostic and hence conventional taxonomy, even though he believed that species were themselves natural, along with genera. This tension underlies much of later taxonomy.

Strictly speaking, Linnaeus did not *have* a "species" concept, despite recent arguments that he held a "biological" species concept [Müller-Wille and Orel 2007]. If he "had" a "concept," it was Ray's generative concept from seed, and merely involving generation is insufficient to make it a reproductive isolation conception of species, which is what "biological" implies in this context.

GEORGES-LOUIS LECLERC, COMTE DE BUFFON:
DEGENERATION, MULES, AND INDIVIDUALS

Georges-Louis Leclerc, Comte de Buffon (1707–1788), referred to simply as Buffon, was one of the last naturalists with an encyclopedic knowledge of zoology. Indeed, he effectively defined that discipline. He was a French aristocrat (a count) who superintended the "King's Garden" *(Le Jardin du Roi,* later *Le Jardin des Plantes).* His pupil and later associate was the famous early evolutionist Lamarck, but Buffon was not what we would understand to be an evolutionist himself.

Buffon strongly disapproved of Linnaeus's binomial system and particularly of his use of sexual characters in discriminating plants. As a result, he and his followers were often in argument and political maneuverings against the Linnaeans. He was the primary author of the forty-four-volume *Natural History, with Particular Reference to the Cabinet of the King (Histoire naturelle,* 1749–1789) of which he issued thirty-six volumes; and in the course of this stylistically elegant but often confusing and sometimes contradictory series, he made a number of passing comments regarding species, which influenced many later ideas on the subject.

Buffon was a relatively standard adherent to the Great Chain—he adopted the "law of continuity" *(lex continua)* of Leibniz and his followers, but he did not necessarily accept the Principle of Plenitude, and so did not expect that every possible kind of species would necessarily exist. He wrote that it was an error in metaphysics trying to find a natural definition of species:

> The error consists in a failure to understand nature's processes *[marche],* which always take place by gradations *[nuances].* . . . It is possible to descend by almost insensible degrees from the most perfect creature to the most formless matter. . . . These imperceptible shadings are the great work of nature; they are to be found not only in the sizes and forms, but also in the movements, the generations and the successions of every species. . . . [Thus] nature, proceeding by unknown gradations, cannot wholly lend herself to these divisions [into genera and species]. . . . There will be found a great number of intermediate species, and of objects belonging half in one class and half in another. Objects of this sort, to which it is impossible to assign a place, necessarily render vain the attempt at a universal system. . . .
> In general, the more one increases the number of one's divisions, in the case of the products of nature, the nearer one comes to the truth; since in reality individuals alone exist in nature. [*Histoire naturelle* (1749), pp. 12, 13, 20, 38; quoted in Lovejoy 1936: 230]

In this opinion, the boundaries between species were arbitrarily drawn—species grade into each other [Farber 1971; Eddy 1994; Roger 1997; Sloan 1979]. However, this was not a statement about transmutation. He merely thought that the variation between species was continuous, not that species continuously arose from prior species. That idea was left to his student Lamarck to elaborate (although Darwin's grandfather Erasmus independently trumped Lamarck by several years with a similar approach). However, Buffon did allow that some change was possible. He thought that there were "types" of organisms roughly equivalent to Linnaeus's genera. The original type was the "true" form, and various species could degrade from that type to become a kind of "monster." Again, the influence of Aristotle is apparent, but there were limits to the sort of change species could undergo.

Two animals are of the same species, he wrote in the second volume of the *Histoire naturelle* [Lovejoy 1959: 93f.],

[I]f, by means of copulation, they can perpetuate themselves and the likeness of the species; and we should regard them as belonging to different species if they are incapable of producing progeny by the same means. Thus the fox will be known to be a different species from the dog if it proves to be a fact that from the mating of a male and female of these two kinds of animals no offspring is born; and even if there should result a hybrid offspring, a sort of mule, this would suffice to prove that fox and dog are not of the same species—inasmuch as this mule would be sterile *[ne produirait rien]*. For we have assumed that, in order that a species might be constituted, there was necessary a continuous, perpetual and unvarying reproduction *[une production continue, perpétuelle, invariable]*—similar, in a word, to that of other animals.

Intriguingly, Mayr [1982: 334] omits the last sentence, perhaps from a desire to find forerunners for his own biological species concept and downplay the morphological aspect of *species* in predecessors. Hence, we again find in Buffon the generative notion of species that, like Ray and Linnaeus and others before and after, includes both form and reproduction.

Buffon had a mechanistic story for what kept organisms reproducing according to type—it involved what he called the *moule intérieure,* or interior mold. This was an epigenetic, but particulate, hereditary factor that was held constant, or at least not deformed too much. It was derived from the *première souche*, the primary stock from which all species of a type degenerated; and because it was shared, he was convinced that all species within the type were actually one species and interfertile, and he undertook experiments to prove this, with limited success.

His was not the first reproductive concept of biological species, but he was first to make reproductive isolation the test of whether two organisms should be included in the same species. At other times he seemed to claim that only the organisms themselves were real and that species were just convenient fictions or names of biologists.[15] Inconsistent in his definitions over the course of the *Natural History*, he had denied that species are real, asserting that only individuals exist (the first example of biological species "nominalism"). Farber [1971] and Eddy [1994] delimit two stages in Buffon's intellectual development. At first, around 1749, in his "Premier discours" (volume 1 of the *Histoire*) he declared that the reality of species cannot be determined using the artificial methods of naturalists. Later, in 1753,[16] he stated that he considered species to be the "constant succession of similar individuals that reproduce" and that "[t]he term *species* is itself an abstraction, which in reality corresponds only to the destruction and renewal of beings through time."

In volume 2 of the *Histoire* (1749), Buffon adopted the notion of the *moule intérieure*; generation was entirely a physical process akin to Newtonian forces, in many ways similar to Darwin's later pangenesis hypothesis [Darwin 1875]. As Eddy recounts the hypothesis, unused molecules in an organism are brought together and reassembled in the form of the organism in the seminal fluids. This notion of form being physical is in itself very Aristotelian, as Sloan [1985] observes, but Buffon is also directly influenced by Leibniz (which Sloan also notes), and so there is a tension in his thought. If species are forms, then forms must be distinct, but if they are arbitrary, and grade into each other, as Leibniz taught the *scala naturae*, then they are not distinct. Eddy [1994: 648n.] discusses this and points out that Buffon did not, in his view, propose at this stage a historical or biological conception of species so much as a logical one, but whether one adopts Sloan's or Eddy's view, he later came to adopt both, and it is in this that he most influenced his pupil Lamarck.

In the period of 1764–1765, Buffon moved to a temporal and physical conception, beyond question. But he did not think that form changed; it was fixed eternally, and at best change could be a process he came to call *degeneration* in an essay in 1766 entitled "De la dégénération des animaux." Form could be modified, he thought, by external environmental changes, but should the "species" be brought back into their ancestral environment, those changes would be reversed [Roger 1997]. Such change from the *première souche* was primarily due to damage; Eddy says that "[f]ar from seeing degeneration in terms of organic history, Buffon saw it as the death of the organic past; through degeneration, organisms lose their organic identity, become weak and vitiated, and in extreme cases, have trouble reproducing" [Eddy 1994: 652]. Roger reports that Buffon carried out experiments on hybridization to test this theory and had surprising success, but eventually Buffon accepted that some animals, particularly domesticated ones, do not revert to the purity of the wild type.

He did, however, allow that the natural unit of biology was not fixed at a single level. What the Linnaeans called genera were roughly more like the natural families Buffon thought set the limits of variety. He noted in volume 6 of the *Histoire* (1779), in the essay on sheep, that human intervention had caused many "species" of domesticated animals to degenerate from a single stock: "These physical genera [of sheep, oxen and dogs] are, in reality, composed of all the species, which, by our management, have been so greatly variegated and changed; as have all those species, so differently modified by the hand of man, have but one common origin in Nature, the whole genus ought to constitute but a single species" ["The Sheep," *Histoire naturelle*, volume 6, 221; quoted in Greene 1959: 147].

In his *Histoire naturelle des oiseaux* (1770), Buffon claims that the size of the species of bird directly correlates with the number of species that degenerate from the *première souche*:

> A sparrow or warbler has perhaps twenty times as many relatives as an ostrich or a turkey; for by the number of relatives I understand the number of related species that are sufficiently alike among themselves to be considered side branches of the same stem, or at least ramifications of stems that grow so closely together that one can suspect that they have a common root, and can assume that originally they all sprang from this root, of which one is reminded by the large number of their shared similarities; and these related species probably have separated only through the influence of climate, food, and the procession of years, which brings into being every realizable combination and allows every possibility of variation, perfection, alteration, and degeneration to become manifest. [Stresemann 1975: 56, translating p. 75 in the original]

Buffon held at the end the view that species were definite categories and that one might determine the boundaries by seeing experimentally whether reproduction was possible; he made much in his essay "Mulets" (1776) of the sterility of hybrids as a mark of species boundaries. But what he meant by *espèces* was more like Linnaeus's genera, or even the later family rank, and he was not fussed about using the Linnaean terminology inconsistently, occasionally shifting from one term to another, perhaps deliberately to annoy the Linnaeans, perhaps doing exactly what Theophrastus had done, using the terms informally. Local variant forms for him were more geographical varieties than they were species, which was the *première souche*. Moreover, he believed that these varieties were closely related to similar forms in the Old as in the New World. It was a kind of vicariance definition of species; Stresemann (1975: 57) gives the example of Buffon putting shrikes (genus *Lanius*) into a single species with species from Senegal *(Tchagra senegala)*, Madagascar *(Leptopterus madagascarinus)*, and Cayenne *(Thamnophilus doliatus)* as climatic variations.

MICHEL ADANSON: MANY CHARACTERS

Michel Adanson was a student and intimate of Antoine-Laurent de Jussieu's uncle, Bernard, and for a time he lived with the de Jussieu family. He laid out a taxonomic procedure that derived from Bernard's ideas in his *Family of Plants* *(Familles des Plantes*, 1763, 1764) in which he attempted a natural system in which all characters are to be used equally to uncover groups. He wrote:

> It was necessary to seek in nature for nature's system, if there really was one. With this aim, I examined plants in all their parts, without omitting one, from roots to embryo, folding of leaves in the bud, anner of sheathing, development, position, and folding of the embryo and radicle in the seed relative to the fruit; in a word, a number of features to which few botanists pay attention. [Morton 1981: 303]

He allowed, where Ray and Linnaeus had not, the use of microscopic characters in classification, as Grew had done. Adanson was later claimed, somewhat illicitly, as a precursor to the so-called phenetics school of classification [Winsor 2004], but it is not the case that he held that all characters have equal weight in classification, but rather that there should be no a priori specification of what characters were to be used, as Linnaeus had done [Stafleu 1963: 184]. He noted that "[w]hat is sufficient to constitute the genera of certain families is not sufficient for other families, and neither the same parts nor the same number of these parts invariably furnish these [constituent] parts in each family" [Morton 1981: 305].

Although he was primarily concerned with higher taxa than the species level, he did give a definition of species in the second part of the preface to the *Familles des plantes*. A definition of species founded only on sexual reproduction fails to apply for all organisms, but only for some plants and most animals. So such a definition is artificial and arbitrary, and if all possibilities are taken into account, the term becomes hard to define:

> Although it is very difficult, not to say impossible, to give an absolute and general definition of any object of natural history whatever, one could say rather exactly that there are as many species as there are different individuals among them, different in any (one or more) respect, constant or not, provided they are definitely perceptible and taken from parts or qualities where those differences appear to be most naturally placed in accordance with the particular character of each family. [translation from Stafleu 1963: 185][17]

He rejected the definition of Buffon, in part because he wanted still to include mineral species under the rubric [Stafleu 1963: 182f.], but also because he recognized the existence of asexual plants and even some animals. He notes that although there may be, as Buffon had said, only individuals in nature and that for God all is one, for us it is divided, and *"et cela sufit"* (and that suffices); the differences are real even if to divine inspection the species are not. He does think, though, that there is continuity in nature and that we see gaps between species only because some have become extinct. In a handwritten definition in the fifth volume of his copy of the *Encyclopédie* of Diderot, he wrote:

> *Species:* collection of all objects which nature separates individually from each other as so many isolated entities existing separately and which the imagination or the free and creative opinion of man unites *idéalement* each time that he finds an almost complete resemblance or a resemblance at any rate greater than with any other group, a collection to which he gives the name species. [translated in Stafleu 1963: 186]

Although it was not published, this passage indicates, as Stafleu notes, that Adanson thought that it was resemblance for the observer that "made" species and that varieties were the subspecific changes occurring in species. So he was a kind of taxonomic essentialist, but was he a species fixist? He sought to answer the question whether natural classification is meaningful if species change, which he believed they had done, and that he had observed it. He cited both Linnaeus's example of *Peloria* and J. Marchant's discovery in 1715 of a mutation in *Mercurialis* [Morton 1981: 309]. His own work with lettuce and basil led him to conclude that varieties had become fixed in each generation. Morton considers that Adanson in fact had a kind of genealogical approach to classification, given that plants were mutable (he is the first to use the term *mutation* in connection with taxonomic transmutation). However, he reexamined the question in 1769 and concluded that the changes he had observed before and since were neither due to hybridism nor did they breed true, and so "the transmutation of species does not take place in plants, any more than in animals." Thus, he ended up a fixist in practice, if not in principle.

ANTOINE-LAURENT DE JUSSIEU: SPECIES AS SIMPLES

Although he was mainly concerned with the nature of a natural system at much higher levels of classification than the species level, Antoine-Laurent de Jussieu (1748–1836) did also make some comments on species that indicate that at the time he published his massive *Genera plantarum secundum ordines naturalis disposita* (1789), two distinct approaches to species had developed. One was due to Ray, while the other was due to Linnaeus, and Jussieu was more in line with Ray than Linnaeus.

Jussieu influenced many, if not most, of those who followed. He directly influenced botany in general—for instance, Gray's *A Natural Arrangement of British Plants* [Gray 1821] shows how he had become the standard even at the popular level. He influenced too the young Cuvier [Stevens 1994: 2, chapter 4] and through him systematics as a whole.

In his early works, he treated species as having more restricted affinity than the affinity of the genus [Stevens 1994: 292; the following quotations from Jussieu's works are from the translations in this book]. He defines species in terms of the sharing of characters [p. 340], noting that "[j]ust so many plants agreeing in all their parts, or being consistent in their universal character, and born from and giving birth to those of like nature, are the individuals together constituting a *species*, [a term] wrongly used in the past, now more correctly defined as *the perennial succession of like individuals, successively reborn by continued generation*" [emphasis added]. So here Jussieu is treating the diagnostic aspect of species, which are formed from individual organisms by abstraction of characters, from the causal aspect of species, which are—again—a generative reality of like producing like, as he does elsewhere [pp. 311, 313–314].

But it is the *Genera plantarum* that had the greatest influence, and in the introduction he explicitly defines species, the concept, thus, in a section entitled "The Sure Knowledge of Species" [p. 356f.]:

> [T]he species must first be known, and defined by its proper signs: [it is] a collection ["adhesio"] of beings that are alike in the highest degree, never

to be divided, but simple by unanimous consent [and] simple by the first and clearest law of Nature, which decrees that *in one species are to be assembled all vegetative beings or individuals that are alike in the highest degree in all their parts, and that are always similar ["conformia"] over a continued series of generations*, so that any individual whatever is the true image of the whole species, past, present, and future. [italics in original][18]

He continues then that genera are analogously formed from species that "conform in a large number of characters" but that there is no hard and fast rule for this in the natural method. Species are the only natural *rank* he recognizes. Species are simples from which the composite groups are formed. Of the Linnaean scheme, he says merely that the nomenclature is useful, a matter of convenience: "the species name must be both simple and easy, but it must also signify" [cf. p. 344], and "It is one thing to name a plant, another to describe it" [p. 510, n. 115].

Like Linnaeus, Jussieu held that plants have affinities like a regions on geographic map. The natural method "links all kinds of plants by an unbroken bond, and proceeds step by step from simple to composite, from the smallest to the largest in a continuous series, as a chain whose links represent so many species or groups of species, or like a geographical map on which species, like districts, are distributed by territories and provinces and kingdoms" [p. 355]. However, as Stevens observes, the geography was becoming sparser in the regions occupied [Stevens 1994: 74ff.]. Whereas for Linnaeus, the entire territory was filled, more or less, for de Jussieu, there were large unoccupied regions, and for Mirbel and de Candolle *fils*, the groupings become discrete and separate.

CHARLES BONNET AND THE IDEAL MORPHOLOGISTS

Charles Bonnet (1720–1793) was a Swiss zoologist who was greatly influenced by Leibniz's ideas about the continuity of nature (the *lex continui*), and he produced the classical ladder of nature as a result, first in his *Traité d'insectologie* (1745) and then in his *Contemplation de la nature* [1764; Stresemann 1975: 172]. His ladder was envisaged as an artificial system of ranking things in terms of their progression, although he did so, he said, without "presuming to establish the progressive order of Nature." The ladder ran from fish to birds to quadrupeds, and each division itself was further divided; birds into aquatic, amphibious, and terrestrial, and so forth (see figure 4).

Bonnet's *Échelle des êtres naturels*, as he called it, implied that extant species were the current forms of "extinct" species but that we can only

Figure 4 Bonner's Scale. Bonner's "Chain of Natural Beings," as presented in his *Contemplation de la nature* (1764), was enormously influential, even though it was a reworking of the medieval notions derived originally from Aristotle. It was a static, rather than temporal, ladder, but Lamarck transformed it into a temporal sequence (see figures 5 and 6). The translation of the scale is as follows:

IDE'E D'UNE ECHELLE DES ETRES NATURELS.

CONCEPT OF A SCALE	OF NATURAL BEINGS
L'HOMME	MAN
Orang-Outang	Orangutan
Singe	Monkey
QUADRUPEDS	QUADRUPEDS
Ecureuil volant	Flying Squirrel
Chauvesouris	Bat
Austruiche	Ostrich
OISEAUX	BIRDS
Oiseaux aquatique	Aquatic Birds
Oiseaux amphibre	Amphibious Birds
Poissons volans	Flying Fish
POISSONS	FISH
Poissons rampans	Crawling (Walking) Fish
Anguilles	Eels
Serpens d'eau	Water Snakes
SERPENS	REPTILES
Limaces	Slugs
Limaçons	Snails
COQUILLAGES	SHELLFISH
Vers à tuyau	Tubular Worms
Teignes	Moths
INSECTES	INSECTS
Gall-insectes	Gall Flies
Taenia, ou Solitaire	Tapeworm
Polypes	Polyps
Orties de Mer	Sea Anemone
[Plantes] Sensitive	Sensitive Plants
PLANTES	PLANTS
Lychens	Lichens
Moisissures	Molds
Champignons, Agarics	Mushrooms, Agarics
Truffes	Truffles
Coraux & Coralloides	Corals & Coraloids
Lithophytes	Lithophyte
Amianthe	Asbestos, Amianthus
Talcs, Gyps, Sélénites	Talcs, Gypsums, Selenites
Ardoises	Slates
PIERRES	STONES
Pierres figurées	Formed Stones (fossils?)
Chrystallisations	Crystals
SELS	SALTS
Vitriols	Vitriols
METAUX	METALS
DEMI-METAUX	SEMIMETALS
SOUFRES	SULPHURS
Bitumes	Bitumens
TERRES	EARTHS
Terre pure	Pure Earth
EAU	WATER
AIR	AIR
FEU	FIRE
Matières plus subtiles	More subtle matters

diagnose the modern forms in terms of their reproduction of like forms. Like the early Buffon, he effectively denied the reality of species considered as essences, and plumped for a nominalistic individualist conception:

If there are no cleavages in nature, it is evident that our classifications are not hers. Those which we form are purely nominal, and we should regard them as means relative to our needs and to the limitations of our knowledge. Intelligences higher than ours perhaps recognize between two individuals which we place in the same species more varieties than we discover between two individuals of widely separated genera. Thus these intelligences see in the scale of our world as many steps as there are individuals. [*Contemplation de la Nature*, 2nd ed., 1769, I, p. 28; quoted in Lovejoy 1936: 231]

Lovejoy notes wryly, "Thus the general habit of thinking in terms of species, as well as the sense of separation of man from the rest of the animal creation, was beginning to break down in the eighteenth century." However, in the same century in which the notion of *species* had acquired a biological sense distinct from the philosophical and logical tradition, it is highly significant that this "nominalism" occurs so early and remains contentious from that day until this.

Donati in 1750 extended this view, says Stresemann, and "therefore conceived the notion that every being is a knot in the web of nature, and its resemblance to other forms may be compared to the threads between the knots" (1975: 172f.), which may have influenced Linnaeus's conception of analogies between orders. Such ideas recur in the work of Johannes Hermann (1783) and Jean Baptiste Robinet (1763), who produced a three-dimensional lattice in which species were nodes in the lattice. Stresemann traces this view through Schelling, Spix, Oken, and others to William Swainson (1975: 174–177). Swainson's view on classification led to a particular account of species as formal ideas. For him, following the ideas of William Macleay published in 1819 [Hull 1988c: 92–96], all taxa had to be organized in circles of affinities that touched ("osculating circles"). Species were arranged at the rim of these circles and were analogous to the species in the adjacent circle. Swainson also had an extended discussion in his *Preliminary Discourse* [Swainson 1834] on what counted as a "natural classification." This debate in part inspired Darwin to think through the reason for systematic classification and the nature of a natural system, and it survived for some time after the *Origin* [Coggon 2002]. In Oken's native Germany, the tradition he had begun continued in the work of Johann Jakob Kaup (1803–1879), who in turn influenced Leopold Fitzinger. Kaup relied on a pentagrammatic system, but otherwise he was much in the tradition of Macleay and Swainson (Stresemann 1975: 178–188). However, such claims of the "end" of morphology depend upon a rather rigid definition of morphological biology, which, as Amundson notes, is neither historically plausible nor accurate.

IMMANUEL KANT AND THE CONTINUITY OF SPECIES

Immanuel Kant's views on biological species are interesting primarily because they influenced the work of Blumenbach [Lenoir 1980], whose own work, published in 1781, established the notion of races as distinct subspecific ranks within the human species [Osborne 1971: 164] and later influenced the *Naturphilosophen* [Amundson 1998; Nyhart 1995]. While Blumenbach worked mainly with skull morphology, his views on teleology were influenced deeply by Kant, and Kant regarded him as the scientist who best understood his ideas. So we may consider Kant's view to be influential on some aspects of his contemporary biology [Moss 2003: 10–12] and also on later biology through Goethe and Oken.

In his early lectures on geology, Kant made a similar distinction to that of Locke, between species defined by similarity relations and genealogical classification, although unlike Locke he specifically named the genealogical groups "natural species" rather than simply "natural relations," as Locke had. He disparages species that are simply similar as "academic" or "scholastic," and as dry, unproductive logical species:

> In the animal kingdom the natural classification into species and variety is grounded in the common law of propagation, and the unity of the species is nothing other than the unity of the generative force, which holds thoroughly for a certain manifold of animals. Accordingly, Buffon's rule that animals which produce fertile offspring with one another, (regardless of the difference in form there may be between them) belong to one and the same physical species, is actually viewed merely as the definition of a natural species of animal in general, in contrast to all scholastic species of animals. The scholastic classification is made according to classes and orders animals according to similarities. The natural classification, however, is based on lines of descent and orders them according to relationships with respect to generation. The former accomplishes a scholastic system for the memory, the latter a natural system for the understanding; the former intends only to bring the creatures under titles, the latter to bring them under laws. [Kant, *On the Different Races of Man*, 1775[19]]

It is unfortunate that Kant did not follow this up further, so far as I know. Natural classification is here based on similar causal processes of fertility and reproduction, a view he based on what turns out to be a misreading of Blumenbach's *Bildungstrieb*, or "formative force" [Richards 2000]. Kant returned to the question of the human races in an essay "Determination of the Concept of Human Races" in 1785, during the course of which he addressed the reproductive aspect of species:[20]

> For animals whose variety is so great that an equal number of separate creations would have been necessary for their existence could indeed belong to a *nominal family grouping* [*Nominalgattung*—literally, nominal species] but never to a *real one*, other than one as to which at least the possibility of descent from a single common pair is to be assumed. . . . [Otherwise] the singular compatibility of the generative forces of *two* species (which, although quite foreign as to origins, yet can be fruitfully mated with each other) would have to be assumed with no other explanation than that nature so pleases. If, in order to demonstrate this latter supposition, one points to animals in which crossing can happen despite the [supposed] difference of their original stems, he will in every case reject the hypothesis and, so much the more because such a fruitful union occurs, infer the unity of the group, as from the crossing of dogs and foxes, etc. The *unfailing inheritance* of peculiarities of both parents is thus the only true and at the same time adequate touchstone of the unity of the group from which they have sprung: namely the original seeds *[Keime]* inherent in this group developing in a succession of generations without which those hereditary variations would not have originated and would presumably not *necessarily* have *become* hereditary. [translated in Greene 1959: 233]

Here we see a beautiful exemplar of the generative conception of species, based on "seeds" with a generative force unique to that kind, which featured prominently in John Ray's initial biological definition of *species*.

The *Critique of Pure Reason* was first published in 1781, the same year as Blumenbach's race work. A second edition[21] followed in 1787, and it is this version [pp. B679–B690] that we will follow here [Kant 1933]. In the context of discussing if a fundamental power exists that unites the things of understanding, Kant asks whether reason derives the unity of things by transcendental employment of understanding—in other words, if parsimony is a law of nature as well as of reason. He answers that unity is a necessity, for otherwise we would have no reason at all, and launches into a standard account of genera and species:

> That the manifold respects in which individual things differ do not exclude identity of species, that the various species must be regarded merely as different determinations of a few genera, and these, in turn, of still higher genera, and so on; in short that we must seek for a certain systematic unity of all possible empirical concepts, in so far as they can be deduced from higher and more general concepts—this is a logical principle, a rule of the Schools, without which there can be no employment of reason. [p. B679f.]

Kant equivocates here, it seems. On the one hand, he wants to adopt the classical process of differentiation from the *summum genera* employed by the Scholastics. On the other, he wants to derive unity from empirical data—that is, to classify from the bottom up. Parsimony is an advance, as when chemists reduce all salts to acids and alkalis [p. B680], "but not content with this, they are unable to banish the thought that behind these varieties there is but one genus, nay, that there might even be a common principle for the earth and the salts" [p. B681]. Parsimony is due to the need for the understanding to reduce multiplicities into unities. This is exactly what the Scholastics would have sought. But he then continues: "The logical principle of genera, which postulates identity, is balanced by another principle, namely that of *species*, which calls for manifoldness and diversity in things, notwithstanding their agreement as coming under the same genus, and which prescribes to the understanding that it attend to the diversity no less than to the identity" [p. B682].

Observation of things must discriminate, he says, as much as the "faculty of wit" must find the appropriate universal. This differs from the Scholastic account, and in some ways is more like Cusa's contraction in species. Here, observation allows us to group diversity of species under general predicates, he says, for if

> there were no *lower* concepts, there could not be *higher* concepts. Now the understanding can have knowledge only through concepts, and therefore, however far it carries the process of division, never through mere intuition, but always again through *lower* concepts. The knowledge of appearances in their complete determination, which is possible only through the understanding, demands an endless progress in the specification of our concepts, and an advance to yet other remaining differences, from which we have made an abstraction in the concept of the species, and still more so in that of the genus. [p. B684]

Kant has it both ways after all: we abstract our more general categories from empirical observation, and understanding divides categories logically so that reason can deal with them [p. B695f.]. But species border each other—there is a logical continuum, "admitting of no transition from one to another *per saltum*, but only through all the smaller degrees of difference between them" [p. B687]. This he calls the logical law of the *continuum specierum*, a version of the transcendental law of *lex continui in natura*, which is Leibniz's law of continuity. While this applies to the realm of possible concepts, though, Kant rejects it in nature: "For in the first place, the species in nature are actually divided, and must therefore constitute a *quantum discretum*. . . . And further, in the second place, we could not make any determinate empirical use of this law, since it instructs us only in quite general terms that we are to seek for grades of affinity, and yields no criterion whatsoever as to how far, and in what manner, we are to prosecute the search for them" [p. B689]. He wants species to be useful in reason and understanding—if there were no gaps in nature, then we could not make sense of it; the fact that things actually are divided by gaps is therefore fortuitous.

There is no biological discussion in this critique, but Kant does provide one in the *Critique of Judgement* in 1790 [second edition in 1793: Kant 1951]. In section 64, he states:

> In order to see that a thing is only possible as a purpose, that is to be forced to seek the causality of its origin, not in the mechanism of nature, but in a cause whose faculty of action is determined through concepts, it is requisite that its form be not possible according to mere natural laws. . . . The contingency of its form in all empirical natural laws in reference to reason affords a ground for regarding its causality as possible only through reason.

It seems Kant is saying that contingent forms (i.e., species) can be understood not as the determinate outcome of mechanisms, but rather as the result of conceptual necessity. Eco discusses this at length [Eco 1999: 89–96] and concludes that the platypus, discovered and displayed in Europe shortly after Kant's death, would have given Kant trouble unless he was able to subsume it under existing conceptual categories (such as "water mole").[22] While the platypus—or to use Kant's own example, a tree—is there and as a natural purpose produces itself as both cause and effect, generically, our ideas of it depend on our knowing the purpose or goal of such organized things [§65]. Kant says in section 67 of the *Judgement*:

> Hence it is only so far as matter is organized that it necessarily carries with it the concept of a natural purpose, because this its specific form is at the same time a product of nature. . . .
> If we have once discovered in nature a faculty of bringing forth products that can only be thought by us in accordance with the concept of final causes, we go further still. We venture to judge that things belong to a system of purposes which yet do not (either in themselves or in their purposive relations) necessitate our seeking for any principle of their possibility beyond the mechanisms of causes working blindly. For the first idea, as concerns its ground, already brings us beyond the world of sense, since the unity of this supersensible principle must be regarded as valid in this way, not merely for certain species of natural beings, but for the whole of nature as a system.

Kant's teleology is too far afield from our topic [see Lenoir 1987], but it is important to see that he saw species as the outcome of self-generative organization in nature, as he had in the lectures of 1775, as well as being things that we needed to think of as goal directed in order to understand them. Again, here is the philosophical current of the generative notion of organic species in play. Famously, Kant was influential on Goethe, and through him, Oken, and the morphologists that followed him (see the later discussion on Owen and Agassiz). In an essay in 1785, he discusses whether all canines are independent creations or descend from a common stock and are races of the one species: "Species and genus [of logic— JSW] are not distinguished in natural history (which has only to do with ancestry and origin). Only in the description of nature, since it is a matter of comparing distinguishing marks, does this distinction come into play. What is species here must there often be called only race" [*Gesammelte Schriften*, VIII, 100n.; quoted in Greene 1959: 372n.]. Here Kant identified the different role the terms *genus* and *species* play in logic and natural history. When we are classifying in nature, we make use of the logic in terms of "identifying marks," but even so we are only often identifying only subspecific groups (races).

WHEN DID ESSENTIALISM BEGIN?

At this point, it might be appropriate to ask when essentialism actually enters the biological debate, since we have not found it so far. There are several senses in which we might call some conception of general terms "essentialist." Locke is an essentialist about terms, for example, since real essences are hidden from us, but terms are just names. Kant is an essentialist about mechanisms, particularly of generation, but in line with his phenomena-noumena distinction suggests that we can only abstract our terms from appearances.

Recently, Amundson suggested that essentialism with respect to species did not begin until the 1840s with the Strickland Code [Amundson 2005], and for biological species this may be true, though I will argue this is merely diagnostic. My argument is that essentialism, construed as the claim that a general term or concept must have necessary and sufficient inclusion criteria, is a long-standing *formal* notion, but when it comes to applying that notion to living things, it was always understood that living species were a different category to formal species.

Let us therefore distinguish, since that is the key to this section, between several senses of "essentialism." We have encountered so far *nominal essentialism* with Locke, the view that names can have essences, but only names. Is Strickland's a nominal essentialism? Not as Amundson presents it. Strickland is more correctly understood to be a *taxonomic essentialism*—that in the process of determining natural groups, one must find what actual properties (in this case, biological properties) they have in common. Taxonomic essentialism is a kind of *logical essentialism*, in that it relies on the construction of formal, or logical, groups, as Aquinas posed it. But it is also a *material essentialism* in Aquinas's terms, because it relies on material properties and not just formal ones. A similar distinction was made by Buridan between formal and material significata, so that he could be a "predicate essentialist" and yet a nominalist [Klima 2005]. This distinction has long standing.

A possible source of the notion that biological species must be essentialist may be John Dewey's famous essay of 1909, on "The Influence of Darwin on Philosophy" [reprinted in Dewey 1997]. Here, Dewey presents a finalistic account of Greek conception of *eidos* (anachronistically stating that the scholastics translated this as *species*, not the Latins). He says that *eidos* keeps the flux stable and makes individuals isolated in space and time keep to a "uniform type of structure and function," and that the conception of species as a "fixed form and final cause" was the central principle of knowledge and nature [pp. 5–6]. But it is fixism and finalism that he opposes to Darwin's theory, not essentialism. His only mention of essence is a sneering reference to "the logic that explained that the extinction of fire by water through the formal essence of aqueousness" [p. 14]; presumably the fault is the use of a formal notion where an efficient cause is required.

Traditional essentialisms are generally nominal. From Aristotle through the end of our period, when people discuss the essences, they are very often discussing what description or definition is essential for a universal name or term. Locke's rejection of Real Essences is a rejection not of the essences of terms but of things. He rejects material essentialism. And it is the material essentialism of biology that is problematic: did it, as a historical fact, occur before Darwin? And is it required for taxonomy? We shall see that neither are necessarily the case, although it is likely that the issues were not so marked as I have expressed them here; and while naturalists do in fact slide from nominal to material essentialisms from time to time, it is not the identifying truth of the period that the Received View/synthetic historiography asserts.

ESSENTIALISM AND NATURAL SYSTEMS

Essentialism and typology are two attributes of "traditional" taxonomy that are often conflated (e.g., by Mayr). But they actually represent two distinct aspects of the old taxonomic categories. Essences are definable, and can be known by refining definitions, and are common to all members of a kind. No member of the kind can *not* have an essential property fully and constantly. On the other hand, types (although the term is used in various ways) are somewhat different. In biological thought, the type of a kind such as a genus can be more or less instantiated[23] and can be varied from. They can be abstract—not actually instantiated in any actual member of the kind, but *every* member of a kind must exhibit essences. Types are formal notions, essences are definitional, and while some types may be essential and some essences may be typical, the two concepts are not identical. Even worse, there are several types of types. Stevens [1994: 134] cites Paul Farber's taxonomy of types [Farber 1976]. According to Farber and Stevens, there are three kinds: collection, classification, and morphological types: "Collection type concepts were concerned with how the name of an organism can be referred to a particular specimen or individual species. Classification type concepts were those that dealt largely with summarizing or simplifying data, whereas morphological concepts dealt with the order of nature and its laws (although Paul Farber, who I am following here, noted that these two were not always sharply distinguishable)." A further discussion of types in the nineteenth century is in Farber's later important paper [Farber 1976]. This is one of the first, if not the first, challenges to the essentialism story.

Divergences from the type of a genus were considered *terata*, or monsters. They were less than perfect and could be individual organisms or even entire species. Hybrids were monsters, too, the sense in which Linnaeus classified his hybrids. In somewhat later thought, such as for Buffon [Gayon 1996; Roger 1997; Stamos 1998], prior to the evolutionary period shortly before the nineteenth century, transmutation was conceived of as degradation from the type. By the time Owen proposed his notion of the *Archetype* that was to influence Darwin [Camardi 2001], the transcendentalists had restored a Platonic view of types as pure forms, as ideas, and as essences [Desmond 1984], but the Aristotelian account allowed only for types as actual forms. In "pure" Aristotelianism, variation from the type was an accidental difference, but as with the Linnaean fixation of the genus-species level, Enlightenment biologists were not always pure Aristotelians. Some of them weren't Aristotelians at all. Linnaeus himself had asserted that "*Natura non facit saltum*," reiterating the principle of plenitude, at least for genera (he crossed this apothegm out in his own copy of the *Philosophia botanica* when he found species doing exactly that in hybrids). This principle owed most to neo-Platonic doctrines of emanation and also relied on the providence and benevolence of God, and it insisted on completeness and a grading from one form to another, as we have seen [Lovejoy 1936].

At the end of the eighteenth century, classifications were commonly thought to be of three sorts: "artificial," "natural," and biological. I put *artificial* and *natural* in quotes because the way these terms are used in the history of systematics is at odds with the meaning of these terms in other fields (and is indeed inconsistent throughout the history of systematics itself).[24] Linnaeus felt that he was promoting an artificial classification— one based on a *single aspect* of organisms (in plants, the sexual system). This was a dichotomous single-key system based on Platonic *diairesis*: each subordinate taxon is distinguished from others in the ordinate taxon by the possession or nonpossession of the key character—winged/nonwinged, two-winged/four-winged, and so forth. This means that many taxa so formed are privative definitions—defined by what they are *not* [Nelson and Platnick 1981]. He did also attempt a fragmentary "natural" system— one that grouped on all available characters—but species remained those groups that shared all of some set of characters. Others, such as Adanson [Croizat 1945; Lawrence 1963; Stafleu 1963], based their classifications on as many characters as could be used, in an attempt to demonstrate natural groups. Arguably, Linnaeus succeeded more than might be expected just from it being an arbitrary system [Cain 1995].[25]

So it is perhaps better if we adopt the distinction between *taxonomic* and other logical essentialisms and *material* essentialism, and see the former as a more or less conventional and harmless aspect of taxonomy, and the latter as something of a rare bugaboo.

I fully concur here with Winsor's [2001] discussion that the myth of essentialism in the history of systematics is largely due to Joseph Cain's 1958 paper "Logic and Memory in Linnaeus' System of Taxonomy," but I wonder also how much Popperian influence predisposed systematists to accept it and spread it. In any case, we have extensively supported Scott Atran's comment, quoted by Winsor: "For my part, I have so far failed to find any natural historian of significance who ever adhered to the strict version of essentialism so often attributed to Aristotle. Nor is any weaker version of the doctrine that has indiscriminately been attributed to Cesalpino, Ray, Tournefort, A.-L. Jussieu, and Cuvier likely to bear up under closer analysis" [Atran 1990: 85].

I am unsure what Atran means by "weaker version"; there is, at least, a diagnostic essentialism, a *taxonomic* essentialism in play with many authors, but only a very few people, and perhaps only Nehemiah Grew, adopted any sort of material essentialism that did not end up collapsing to a causal reproductive account, prior to the *Origin of Species*. And when it existed, before or after, it was motivated by piety rather than science. The transcendentalists, however, appear to be motivated also by a neo-Platonic philosophy.

THE ORIGINS OF SPECIES FIXISM

Species fixism, the idea that species are as they have always been, appears to have originated with Ray, and it was the standard opinion in popular botanical texts such as Gray's *Natural Relations of British Plants* of 1821, in which "race" is defined as either "primitive" *(plantae primigeniae)*—"Species originally created, and not formed by crossing with others"—or as "mule" *(hybridae)*—"Species not originally created, but formed by the pollen of one species being absorbed by the female organ of another species" [Volume I, p. 41]. Note that no requirement that the hybrids are infertile is assumed here.

That species were in some manner *mutable* (which is perhaps how we should contrast fixism, not with the later transmutationism that followed Lamarck) during the period up until the seventeenth century is held to have been the standard view [Zirkle 1959]. Indications that species were not commonly thought to be fixed in that period can be found in the work of one of the translators of the King James Bible (1611), the Calvinist George Abbot (1562–1633), Archbishop of Canterbury. In his *A Briefe Description of the Whole World* (1605), he wrote:

> There be other Countries in *Africke*, as *Agtsimba* [?], *Libia interior*, *Nubia*, and others, of whom nothing is famous: but this may be said of *Africke* in generall, that it bringeth forth store of all sorts of wild Beasts, as Elephants, Lyons, Panthers, Tygers, and the like: yea, according to the Proverbe, *Africa semper aliquid apportat novi*; Often times new and strange shapes of Beasts are brought foorth there: the reason whereof is, that the Countrie being hott and full of Wildernesses, which haue in them little water, the Beastes of all sortes are enforced to meete at those few watering places that be, where often times contrary kinds haue conjunction the one with the other: so that there ariseth new kinds of species, which taketh part of both. [Nicolson 2003: 160][26]

It is noteworthy, though, that novel species are formed from hybridization of extant, and presumably created, species. Thomas Huxley and others thought that fixism is due to the work of John Milton [Huxley 1893b], but Milton's culpability is hard to determine. In *Paradise Lost*, Book VII, the creation story of Genesis is repeated with little change, and the term used there is *kind*, as it is in the English Bible. Nowhere in his poetic works can I find a hint of the constancy or otherwise of species.

Zirkle [1959: 642] doesn't deal with the origin of fixism except to say, "The idea of the complete fixity of species was beginning to take shape in other quarters [than natural history—JSW], however, and it had become the accepted belief of the theologians. The theologians now held that species remained just as God had made them in the six days of creation—they remained just as God wanted them to be."

But I would cavil at his claims that species were therefore mutable before Ray—it is not that species were not held to be fixed before the seventeenth century; they simply had no *idea* of a biological species. Remember that in the Latin tradition, *genus* means a general kind, while *species* means a special kind. Taking a hint from Locke and replacing *genus* with "kind" and *species* with "sort" for anyone before Ray, we can see that the problem lay in the ways things were broken down into kinds and sorts. Of course, if you graft a branch from an apple tree onto a pear tree, you have a different "sort" of fruit (examples Zirkle gives from Theophrastus). That doesn't mean it's a new *biological* kind or sort, but a new *practical* one.

Of course, a lot of the mutability of species/sorts is also a claim of biological species (in our sense) mutability. There are three kinds in the pre- and early scientific literature that I have so far found:

Mutability from hybridization. Two species can interbreed to form a novel one.

Spontaneous generation or transmutation of one form (say, a worm) into another (say, a goose). This might be seen in modern terms as a part of the single species' life cycle.

New varieties. A species can breed untrue in some characters (even Aristotle knew this); by extension you can get new species. This is almost always put down to the influences of the local country—the soil, climate, or water—that directly changes the breed.

Only the last is relevant for us, and even here I have my doubts that many people thought there would be much change beyond the modern genus level, if that. Amundson [2005, §2.2] has argued that fixism was invented in the mid–seventeenth century. I think he is right, although I suspect Ray, not Linnaeus, is the culprit. But no matter who invented it, the question is why. The answer is, as Amundson shows, that fixism was an outcome of the generation debates [Gasking 1967]. Spontaneous generation [Farley 1977], a view that went back to Aristotle and earlier [Lennox 1981], did imply species mutability, since the generated form could change into a quite different form, as we saw with the barnacle goose case. When the debate over generation resolved in favor of epigenesis rather than preformationism, it followed that the generation of an organism had to be controlled in some manner [Gasking 1967: 34]. Even the preformationists held that something made things develop according to their kinds—they called it the *emboîtement*, and Gasking notes [p. 42] that

> the seventeenth century preformationists assumed . . . that all living things there were to be, had in fact been organized at God at creation . . . when all the created germs had reached the adult form the species would become extinct. . . . [I]t followed from such a view that there was no true generation; what appears as the formation of a new individual was simply the growth of an organised living thing which had been formed at the beginning of Time.

Gasking mentions that Ray was a preformationist, although he rejected Leeuwenhoek's sperm-based version, in favor of the egg-based one that was more traditional [pp. 43, 56; see also Raven 1953].

The kind of mutabilism that existed immediately prior to Lamarck and Erasmus Darwin, excepting that of Pierre Maupertuis, was the kind that Linnaeus allowed—species could be formed by hybridization of existing species. God, of course, formed the original stock, but new forms could arise, especially in plants, by mixing them. There was no open-ended mutabilism. However, it is equally clear that essentialism is not directly tied into the origination of fixism.

Amundson rightly argues that fixism was a precondition for the Natural System, which was itself a precondition for evolutionary theory. But that is not, of course, why it was adopted. Linnaeus did not seek to establish a natural system (not in the *Systema naturae*, at any rate), although his scheme came to be known later as *the* Natural System [Smith 1821], as a result of its title. His fixism was probably more a function of his piety than of his taxonomic concerns. So it may be that fixism was successful simply because it was consonant with the tenor of the times and because it came along with Linnaeus's success in establishing taxonomic nomenclature. Or it may be that Ray's institution of the natural theology movement was the more important, seeing nature as the book in which God had written. Either way, fixism comes into prominence around the end of the seventeenth century, and not in the time of Aristotle.

But we should be wary of papers like Zirkle's and claims that mutabilism was rife before Linnaeus or Ray. In fact, lacking the basic conception of a biological kind (different to any other kind) and still hazy on the mechanisms and behaviors of generation (including spontaneous generation), the pre-Linnaeans tended to use the kind terms *genus, species, varietas,* and *formas* informally, as we would use *sort, kind, variety,* and *form.* Grand conclusions cannot be drawn from this.

THE EARLY NINETEENTH CENTURY: A PERIOD OF CHANGE

> The ordinary naturalist is not sufficiently aware that when dogmatizing on what species are, he is grappling with the whole question of the organic world & its connection with the time past & with Man; that it involves the question of Man & his relation to the brutes, of instinct, intelligence & reason, of Creation, transmutation & progressive improvement or development. Each set of geological questions & of ethnological & zoo. & botan. are parts of the great problem which is always assuming a new aspect.
>
> **Charles Lyell, February 11, 1857 [Wilson 1970: 164]**[1]

NINETEENTH-CENTURY LOGIC

Early in the nineteenth century, in 1826, Archbishop Richard Whately published an influential textbook on logic, *The Elements of Logic* [1875, my edition being the ninth], which is credited as reviving the study of logic in English-speaking countries. In this book, Whately describes *species* as *essences*, as *heads of predicables*, and as that of which genera are parts [and not species being parts of genera, since the genus partakes of the essence, or definition, of the species; Book II, chapter 5, §3, p. 85]. But he also notes that *this* sense of "species" is quite distinct from the sense in which *naturalists* use it of "organized beings"[2] [Book IV, chapter 5, §1], for they are real things, "unalterable and independent of our thoughts" [p. 183[3]]:

> [I]f anyone utters such a proposition as . . . "Argus was a mastiff," to what head of Predicables would such a Predicate be referred? Surely our logical

principles would lead us to answer, that it is the *Species*; since it could hardly be called an Accident, and is manifestly no other Predicable. And yet every Naturalist would at once pronounce that Mastiff, is no distinct Species, but is only a *variety* of the Species Dog. . . .

[T]he solution of the difficulty is to be found in the peculiar technical sense . . . of the word "Species" when applied to *organized Beings*: in which case it is always applied (when we are speaking strictly, as naturalists) to individuals as are supposed to be *descended from a common stock*, or which *might* have so descended; *viz.* which resemble one another (to use M. Cuvier's expression) as much as those of the same stock do.

Whately expressly exempts species concepts in biology, then, from the strictures of logical notions, and that, it must be observed, includes essential characters. He notes. "[The fact of two organisms being the same species] being one which can seldom be *directly* known, the consequence is, that the *marks* by which any Species of Animal or Plant is *known*, are not the very *Differentia* which *constitutes* that Species" [p. 184f.]. So well prior to Darwin, and in a logical context, we find that the *species* of biology and the *species* of logic are understood to be different concepts. However, Whately expects there *will* be diagnostic "marks."

A critical review of Whately's logic entitled *An Outline of a New System of Logic* was published the next year by George Bentham [1827], who later became a noted botanist and who was the nephew of and influenced by his uncle Jeremy Bentham. Bentham attacked Whately for not allowing privative classifications; oddly he thought that was required by Aristotle, rather than by the later neo-Platonists. However, Bentham made a crucial distinction that matches Aquinas's material-formal distinction. While allowing that the differentia of a (logical) species was necessary to it and that if the species is a universal, then it is its essence [p. 67], "without which the subject would not be what it is said to be" [p. 68], he noted also that "the peculiar sense in which naturalists make use of the word species . . . is very different from the logical sense of the word" [p. 71], and he distinguished between the *definition* of a species from its *individuation* [p. 79], in which "the only characteristic properties are those of *time* and *place*, which must both be exhibited," and essential definition applies only in the first case. Furthermore, he notes that the object of specific description is to enable a learner to recognize a species or fix the collective entity in his mind [p. 82]. Description should thus be preceded by definition. However, he equivocates when discussing essential properties and says that "*to the genus plant belong all those entities which are endowed with the property of possessing* leaves, stalks, roots, &c." Even so, he realizes that classification, which he calls *methodization* is either physical or logical [p. 98] and that division is of several kinds—analytic when performed on individuals, logical when performed on collective entities, and possibly also physical. There is a difference between dividing, say, Vertebrata into mammals, birds, fishes, lizards, and serpents in terms of obvious features if you want a "slight and general idea" of vertebrates, but "it cannot suffice for the naturalist, who must always be assured of the all-comprehensiveness of his classes and subclasses; he must always be enabled to ascertain precisely to *which* of them he should refer any individual animal that comes under his observation" [p. 112]. So he applies a binary dichotomy of lungs/no lungs (fish), with mammaries/not with mammaries (mammals), winged (birds)/not winged (reptiles) [p. 114]. In the rest of his work, he seems to have reinvented the Universal Language Project. In his 1817 *Chrestomathia* (Bentham 1983), Jeremy Bentham revives, explicitly (Table IV in that work), Porphyry's dichotomous ("bifurcating") mode of classification, counterpointing it to D'Alembert's classification of knowledge in the *Encyclopedia*. His division of knowledge (his Table V) begins with "Eudaemonics, or Ontology," which he divides into "Coenoscopic" and "Idioscopic" ontologies (general and particular properties), and so on, but lower down he starts introducing privative categories, such as "no-work-producing," "not-state-producing" and so on. And some of his dichotomies, such as Nature/Man and information/passion seem as arbitrary as Plato's original. Still, Bentham's recasting of the Arbor Porphyriana as a bifurcating tree diagram is significant.[4]

By the middle of the nineteenth century, despite the arguments naturalists were now having over the meaning of the term *species*, the "genera plus differentia" definition remained widely accepted by logicians until, under the weight of the new set theory and the biological preeminence of the use of the term, the older logic was relegated to specialists in metaphysics and medievalists. Here, for example, is the definition of a widely used dictionary of science and the arts in 1852:

SPE'CIES. (Lat.) In Logic, a predicable which is considered as expressing the whole essence of the individuals of which it is affirmed. The essence of an individual is said to consist of two parts: 1. The material part, or genus; 2. The formal or distinctive part, or difference. The genus and difference together make up, in logical language, the species: e.g. a "biped" is compounded of the genus "animal," and the difference "having two legs." It is obvious that the names *species* and *genus* are merely relative; and that the same common terms may, in once case, be the species which is predicated of an individual, and, in another case the individual of which a species is predicated: e.g. the individual, Cæsar, belongs to the species man;

but man, again, may be said to belong to the species animal, &c., as we contemplate higher and more comprehensive terms. A species, in short, when predicated of individuals, stands in the same relation to them as the genus to the species; and when predicated of other lower species, it is then, in respect of these, a genus, while it is a species in respect of a higher genus. Such a term is called a subaltern species or genus; while the highest term of all, of which nothing can be predicated, is the "summum genus;" the lowest of all, which can be predicated of nothing, the "infimæ species." The difference which, together with the genus, makes up the species, is termed the "specific difference." [Brande and Cauvin 1853: 1137]

By the third edition in 1859, the discussion had been rewritten by Richard Owen to include the biological meaning of Cuvier, but this is as succinct a summary of the traditional conception as one will find. Moreover, we should note that it follows Aristotle in rejecting binary diairesis in favor of multiple species per genus, each of which carries its own special differentiae ("specific differences").

However, this view was not necessarily the view held by the leading philosophers of the day. Mill and Whewell in particular had tried to accommodate the current facts of natural history into the notion of a classification [Hull 2003]. In his System of Logic of 1843 [Mill 1930], Mill showed considerable knowledge of botanical classification conventions and awareness of variation within species. The discussion in Book I, chapter 7, especially sections 3–4, is well informed as to scholastic and biological conventions, and attempts a reconciliation of the two, without much success.

In Book I, chapter 8, section 4, he discusses the Cuvierian use of the term Man as the scientific definition: "Man is a mammiferous animal having two hands." This defines Man by giving "the place which the species ought to occupy in that particular [scientific] classification." He notes the Aristotelian use of per genus et differentiam, which he seems not to challenge. It is significant not for its resolution of the topic but because we see here a philosopher taking pains to use as many biological examples as possible, although we also see elements and minerals appearing in the exempla gratia. He defines species, at least in the sense used by naturalists [Book I, chapter 7, §3], as "not, of course, the class in the sense of each individual of the class, but the individuals collectively, considered as an aggregate whole." Mill clearly is treating the species of the naturalist in a different sense, a "popular acceptation" more general and less logical than the sense of the logician. Nevertheless, both he and Whewell treated species as "natural classes," as Whewell stated it [Hull 2003: 184f.], and in his response to Darwin's Origin, Whewell was dismissive of the idea that these classes could change. Hull quotes him as saying that "a natural class is neither more nor less than the observed steady association of certain properties, structures, and analogies, in several species and genera" [Whewell 1831; quoted in Hull 2003: 185]. Whewell's Humean associationist psychology is evident here, but also the Lockean idea of general terms as creations of the mind to collate past experiences. It is unclear if this was, as Hull suggests, the core of his objections to Darwin's theory of evolution, but at the least his philosophical adherence to logical essentialism certainly played a part in it.

Interestingly, *after* Darwin, a kind of essentialism regarding species was espoused by Jevons [1878: 710–713], but he makes it clear that this refers to diagnostic species—that is, to classes of definitions of objects. He states that in a "natural" system of classification, all "arrangements which serve any purpose at all must be more or less natural, because, if closely enough scrutinised, they will involve more resemblances than those whereby the class was defined" [p. 680] and thus they are inductive groups, based, in living beings, on "inherited resemblances," such that the "*arrangement . . . would display the genealogical descent of every form from the original life germ*" (italics in original). Therefore, diagnostic essences are correlations that are causally important. Jevons follows Porphyry in treating Species as a predicable [p. 698]. Sir William Hamilton, however, in his *Lectures on Metaphysics* [Hamilton et al. 1874, vol. 2, Lecture XXXVII], notes that we begin classification in "vague and confused" generalities, from which we refine our discriminations of things until we end, rather than commence, with the individuals, so that the genealogy of our knowledge is rather the history of how we came to know them than the history of how they came to be. Species are a product of our getting to know things [p. 334], giving an analogy:

> We perceive an object approaching from a distance. At first we do not know whether it be a living or an inanimate thing. By degrees we become aware that it is an animal, but of what kind—whether man or beast—we are as yet not able to determine. It continues to advance, we discover it to be a quadruped, but of what species we cannot yet say. At length, we perceive it is a horse, and again, after a season, we find that it is Bucephalus.

The use of the notions of *genus* and *species* in logical discussions seems to have petered out with the introduction of set theory and formal logic by Venn, Cantor, Peirce, Frege and others toward the end of the nineteenth century, especially around 1870–1878 in the case of Cantor. Where the inclusion of a species in a genus and of lower species in that species was the mainstay of classificatory logic prior to this development, now the talk was of sets and subsets. Moreover, the introduction of set theory itself seemed to override the older approach of *diairesis*, or top-down division. Sets could be defined from larger sets by division or by aggregation of smaller sets. Even more radical was the distinction between intensional and extensional definitions of sets.[5] A species in the older logic had to be definable from the larger genus. A set could be described *or* defined. An extensionally defined set is treated as isomorphic with another set, or, as Quine [1970: 67] expresses it, "the *law of extensionality*, which identifies sets whose members are the same"; intensions (which Quine abhors) are specified by predicates, which "have attributes as their 'intensions' or meanings." Under the Aristotelian account, all species were intensionally defined—this was the point of defining them by their essences. Now we had aggregates that could be treated as synonymous merely by sharing all members, irrespective of their essences. It was not immediately clear how this might apply to the biological species problem, and it indeed took some time for it to be applied.

Similar essentialist accounts of biological species are also presented after Darwin by several Roman Catholic authors, including a respected entomologist [Wasmann 1910] and a logician [Clarke 1895], both Jesuits. A late example is a Dominican, Murray [1955], and a well-known Canadian entomologist, William Thompson, is another [Thompson 1958, 1971; Thorpe 1973]. Thompson is particularly interesting as he explicitly based his attack on neo-Thomist philosophy. One might conjecture that the essentialism of biological species bemoaned by Mayr and others is in fact a *reaction* to Darwin and evolution (or at least Haeckel's Romantic version of it) rather than something he overcame, as the Received View has it. Perhaps, like Donne, they found all coherence gone with a temporal and gradual transmutation of species one into another. Perhaps it was the outworking of the revival, or rather invention, of neo-Thomism after the First Vatican Council.[6] Amundson, however, has argued that essentialism was invented in the 1840s by Hugh Edward Strickland, but his was a purely taxonomic essentialism with no causal or empirical consequences [Amundson 2005: 51].

Even so, some continued to use the older logical terminology and the Aristotelian conceptions that underlay it well into the twentieth century, even if there were some concessions to the new set theory. Husserl, for example, writing in 1913 [Husserl 1931], writes in section 12 of his *Ideas*:

> Every essence, whether it has content or is empty (and therefore purely logical), has its proper place in a graded series of essences, in a graded series

of *generality* and *specificity*. The series necessarily possesses two limits that never coalesce. Moving downwards, we reach the *lowest specific differences* or, as we also say, the *eidetic singularities*; and we move upwards through the essences of genus and species to a *highest genus*. Eidetic singularities are essences, which indeed have necessarily "more general" essences as their genera, but no further specifications in relation to which they themselves might be genera (proximate or mediate, higher genera). Likewise that genus is the highest which no longer has any genus above it.

More interestingly, and influentially on the subsequent debate, H. W. B. Joseph's *Introduction to Logic* (1st ed. 1906; 2nd, 1916) allowed that the evolutionary species of Darwin and Spencer were a different notion to that of the logical species of definitions. He goes so far as to note that species in biology cannot be defined and that instead one must describe a type, from which individuals can diverge:

The difficulty of determining what attributes are essential to a substance, and therefore of discriminating between essence and property, does not however arise entirely from the seeming disconnexion of the attributes of a kind. It arises also, in the case at least of the organic, from the great variation to which a species is liable in divers individuals. Extreme instances of such variation are sometimes known as border varieties, or border specimens; and these border varieties give great trouble to naturalists, when they endeavour to arrange all individuals in a number of mutually exclusive species. For a long time the doctrine of the fixity of species, supported as well by the authority of Aristotle and of Genesis, as by the lack of evidence for any other theory, encouraged men to hope that there was a stable character common to all members of a species, and untouched by variation; and the strangest deviations from the type, excluded under the title of monstrosities or unnatural births, were not allowed to disturb the symmetry of the theory. Moreover, a working test by which to determine whether individuals were of different species, was furnished, as is well known, by the fertility of offspring; it being assumed that a cross between different species would always be infertile, as in the case of the mule, and that when a cross was uniformly infertile, the species were different. But now that the theory of organic evolution has reduced the distinction between varietal and specific difference to one of degree, the task of settling what is the essence of a species becomes theoretically impossible. . . .

If [biological] species were fixed: if there were in each a certain nucleus of characters, that must belong to the members of any species either not at all or all in all: if it were only upon condition of exhibiting at least such a specific nucleus of characters that the functions of life could go on in the individual at all; then this nucleus would form the essence of the kind. But such is not the case. The conformity of an individual to the type of a particular species depends on the fulfilment of an infinity of conditions, and implies the exhibition of an infinity of correlated peculiarites, structural and functional, many of which, so far as we can see . . . have no connexion one with another. There may be deviation from the type, to a greater or less degree, in endless directions; and we cannot fix by any hard and fast rule the amount of deviation consistent with being of the species. . . . Hence for definition, such as we have it in geometry, we must substitute classification. . . . A classification attempts to establish types. [Joseph 1916: 81f., 88f.]

Joseph continues in the tradition of Whately, separating logical species defined by essence and biological species described by types. Even at this late stage, types and essences are explicitly held to be different notions.

Until Woodger introduces symbolic logic to biology [Woodger 1937, 1952], such issues are discussed by a declining number of philosophers—and then, of course, in the context of both cladism and the individuality thesis.

JEAN BAPTISTE DE LAMARCK: UNREAL SPECIES CHANGE

No sooner had natural history established a tradition of fixism of species than it was immediately under challenge, for example, by Pierre Maupertuis in *Vénus Physique* in 1745. At the turn of the nineteenth century, there was a considerable amount of ferment over the notions of taxonomic groups or ranks. For example, Blumenbach had classified the human species into races—Caucasian, Mongolian, Ethiopian, American, and Malayan—yet he still regarded these types as subordinate to the human species and believed that all were varieties of that species [Nordenskiöld 1929: 306; Voegelin 1998], although Buffon had previously denied that the notion of "race" applied to the usual human groupings [Roger 1997]. Blumenbach's conception of the species was that it was formed through the action of a formative force, a *nisus formativus*, and so his is also a generative notion of species.

More influentially, Lamarck delivered a transmutationist view of species and followed his mentor Buffon in supposing that there were no realities attaching to the term. In the *Zoological Philosophy* [Lamarck 1809; English translation, Lamarck 1914], he writes:

It is not a futile purpose to decide definitely what we mean by the so-called *species* among living bodies, and to enquire if it is true that species are of absolutely constancy, as old as nature, and have all existed from the beginning just as we see them to-day; or if as a result of changes in their environment, albeit extremely slow, they have not in the course of time changed their characters and shape. . . .
Let us first see what is meant by the name of species.

Any collection of like individuals which were produced by others similar to themselves is called a species.
This definition is exact: for every individual possessing life always resembles very closely those from which it sprang; but to this definition is added the allegation that the individuals composing a species never vary in their specific characters, and consequently that species have an absolute constancy in nature.
It is just this allegation that I propose to attack, since clear proofs drawn from observation show that it is ill-founded. [Lamarck 1914: 35]

As Gillispie [1959: 271] puts it, "[Lamarck's] position is rather that species do not exist, than that they are mutable." Lamarck reiterates in the *Zoological Philosophy* the early view of Buffon that only individual organisms exist in nature: "Thus, among living bodies, nature, as I have already said, definitely contains nothing but individuals which succeed one another by reproduction and spring from one another; but the species among them have only a relative constancy and are only invariable temporarily" [p. 44].

It is interesting to note that here and elsewhere Lamarck explicitly restricts his comments to living bodies. His nominalism with respect to organisms is obvious, and species are not themselves in nature, but, as Locke

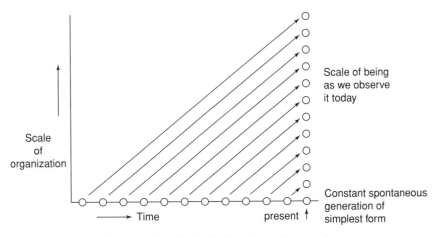

Figure 5 Lamarck's view of evolution had each species at a given moment as the present level of organization achieved by a lineage that was not commonly descended from any prior lineage. Lamarck's conception of species necessitated that the entire species evolved simultaneously [redrawn from Bowler 1989a: 85; a similar diagram can be found in Panchen 1992: 60]. This diagram was devised in M. P. Winsor's PhD dissertation (Winsor, personal communication) and shows Lamarckian evolution in the absence of deviation from the course of nature. Figure 6 shows the deflection that has occurred, in Lamarck's understanding.

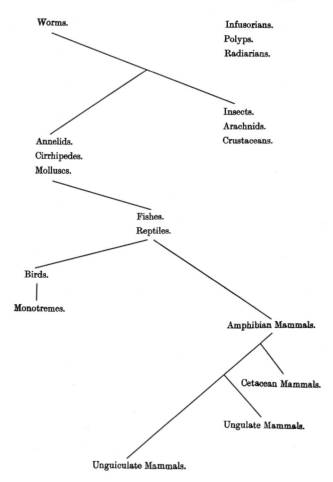

A   TABLE

SHOWING THE ORIGIN OF THE VARIOUS ANIMALS.

Worms.

Infusorians.
Polyps.
Radiarians.

Insects.
Arachnids.
Crustaceans.

Annelids.
Cirrhipedes.
Molluscs.

Fishes.
Reptiles.

Birds.

Monotremes.

Amphibian Mammals.

Cetacean Mammals.

Ungulate Mammals.

Unguiculate Mammals.

Figure 6 Lamarck's scale. "Table: showing the origin of the various animals." A) the 1914 English version. B) Lamarck's original 1809 diagram. The ladder that Lamarck adopted [Lamarck 1914: 179], however, was less direct than Bonnet's. He wrote, "I do not mean that existing animals form a very simple series, regularly graded throughout; but I do mean they form a branching series, irregularly graded and free from discontinuity, or at least once free from it" [p. 37].

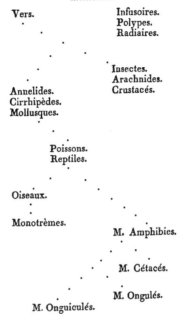

B TABLEAU
Servant à montrer l'origine des différens animaux.

Vers.

Infusoires.
Polypes.
Radiaires.

Insectes.
Arachnides.
Crustacés.

Annelides.
Cirrhipèdes.
Mollusques.

Poissons.
Reptiles.

Oiseaux.

Monotrèmes.

M. Amphibies.

M. Cétacés.

M. Ongulés.

M. Onguiculés.

had said, are made for communication: "Nevertheless, to facilitate the study and knowledge of so many different bodies it is useful to give the name species to any collection of like individuals perpetuated by reproduction without change, so long as their environment does not alter enough to cause variations in their habits, character and shape" [Lamarck 1914].

The generative conception is again in play, except that Lamarck has added temporality to the mix. In the *Recherches* [Lamarck 1802], Lamarck proposed that there was a "life-fluid" that was a variety of physical energy, a *feu éthéré*, that maintained organisms in their form, and it impelled spontaneous generation out of inanimate matter. Species was a notion that applied to the mineral kingdom as well as the biological, but mineral species differed in that they had no individuality, while plants and animals did, and neither did they reproduce. In the *Zoological Philosophy*, he added that *all* classifications are arbitrary products of thought and that in nature there are only individuals. Lamarck did not accept the reality of extinction apart from human agency, but he held that fossil species merely transformed into later forms [Nordenskiöld

1929: 325]. Gillispie notes that Lamarck was slightly inconsistent on species between 1797 and the statements of 1802 and 1807, but he goes on to say [p. 272], "[T]he inconsistency on species [is] trivial. . . . All he did between 1797 and 1800 was to assimilate the question of animal species—or rather their nonexistence—to that of species in general. For in Lamarck the word has not lost its broader connotations. It still carries the sense of all the forms into which nature casts her manifold productions in all three kingdoms (or rather in both divisions)."

Lamarck is still indebted to the medieval notion of species as subsidiary divisions of the *summum genus* (being) through to the infimae species of rational living things in the case of humans. In this respect, he was attempting to classify all things as the outcome of physical molecules and forces, and animal species were just an arbitrary part of that chain of being. Still, his view of these nominalistic species is that they are formed out of the generative properties of the life-fluid, and so this is a generative notion of species in that respect at least.

As to nomenclature, he accepted Linnaeus's binomial convention and, given there was no fact of the matter, held that an international agreement should be made to make names stable [Nordenskiöld 1929: 326]. At this stage, Buffon's objections to the binomial nomenclature have lost the field entirely, when even his own student accepts the practice. Lamarck's view of evolution (figure 5) is basically a temporalization of the ladder of nature/Great Chain of Being [Jordanova 1984; Bowler 1989a; Lovejoy 1936]. He treated each species as a single lineage that had its own original spontaneous generation out of nonliving material and that ascended something like Bonnet's ladder, although the ladder could branch, as we see in the famous diagram (figures 6a and b).

Lamarck's definition was echoed by the son of his defender Geoffroy Saint-Hilaire (1772–1844), Isidore Geoffroy Saint-Hilaire (1805–1861), in 1859, in his *Histoire naturelle générale des règnes organiques* [vol. 2, 437, quoted in Lherminer and Solignac 2000: 156]: "The species is a collection or a succession of individuals characterized by a whole of distinctive features whose transmission is natural, regular and indefinite in the current order of things."[7]

CUVIER: FIXED FORMS AND CATASTROPHES

Lamarck's nemesis, Georges Leopold Chrétien Frédéric Dagobert, Baron Cuvier (1769–1832), was by contrast a full-blooded species realist. Not only were species real, they did not transmute from one to another. Species were generated in some manner at the time of a catastrophe, and previous forms were obliterated. Nordenskiöld writes:

> The immutability of species is to Cuvier's mind an absolute fact; he has not a trace of Linnaeus's hesitation, which he expressed in his old age, in face of the difficulty of drawing a line of demarcation between the species; according to Cuvier's definition, species consist of "those individuals that originate from one another or from common parents and those which resemble them as much as one another." In this definition no mention is made of the creation of the species, which, it will be remembered, Linnaeus took as his starting point, but which, on the whole, Cuvier does not discuss at all. [Nordenskiöld 1929: 339]

Cuvier's definition [found in his 1812 work][8] is interesting in several respects. Despite the superficial resemblance to Linnaeus's definition given earlier, Cuvier's more closely resembles John Ray's definition. It is a historical definition, yet it requires resemblance, presumably to bar monsters. It is, as was Ray's, a generative and yet still a formalist definition. For Cuvier, species come into existence at the beginning of each geological epoch and never vary thereafter. He notes that if species have changed by degrees, then we ought to have found traces of these gradual modifications and intermediate forms between, say, a paleotherium and modern elephant, which we do not. Hence, species are stable. Although spare, Cuvier's definition was very influential on philosophers, as the discussion in Whately's and Mill's logics show. His fixism has the following rationale: "These forms are neither produced nor do they change of themselves; life presupposes their existence, for it cannot change except in organisations ready prepared for it" [p. 20, quoted in Russell 1982].

In his *Éloge*, or funeral oration, of Lamarck [Cuvier 1835], Cuvier set to demolishing Lamarck's idea of species transmuting. Basing his argument on the static nature of Egyptian mummies of various animals when compared with the modern version, and the lack of apparent progress from simple to complex forms that Lamarck's view required in the fossil record, Cuvier established the default view that species did not themselves change. He did not require, as Russell notes [1982: 43] that faunal epochs were an illusion—he simply had no theory of how the new faunas came into existence, as he had little time for theory without facts. He tried to minimize the number of special creations, saying, "I do not pretend that a new creation was required for calling our present races of animals into existence. I only urge that they did not anciently occupy the same places, and that they must have come from some other part of the globe" [*Essay*, 1818 English translation, p. 128; quoted in Greene 1959: 363n.].

JAMES PRICHARD: SPECIES ARE REAL; VARIATIONS ARE ENVIRONMENTAL

James Prichard (1786–1848) published his widely read *Researches into the Physical History of Man* in the same year Cuvier's "Discours" was published [Prichard 1813]. In it, he foreswore a priori argument, whether scriptural or not, and vowed to deal only with evidence, not speculation. That notwithstanding, Prichard declared that Linnaeus's classes were arbitrary and artificial (something we have seen Linnaeus might admit freely), but

[n]ot so in the case of species. Here the distinction is formed by nature, and the definition must be constant and uniform, or it is of no sort of value. It must coincide with Nature.

Providence has distributed the animated world into a number of distinct species, and has ordained that each shall multiply according to its kind, and propagate the stock to perpetuity, none of them ever transgressing their own limits, or approximating in any great degree to others, or ever in any case passing into each other. Such confusion is contrary to the established order of Nature.

The principle therefore of the distinction of species is constant and perpetual difference. [pp. 7–8, quoted in Greene 1959: 239]

He then used this criterion to reject Buffon's concept of species and to argue that human races were all of the same species. He denied that local variations were caused by the direct action of the soil and environment, arguing that variation occurred naturally when the environment changed and was passed on (in the context of human variation). Civilization minimized this variability in humans, and so dark skins, due to savage conditions, would turn into light skins upon civilizing.

Prichard's view may have influenced the well-known view of William Whewell in 1837 that "*species have a real existence in nature*, and a transition from one to the other does not exist" [Whewell 1837, vol. 3, p. 626, quoted in Hull 1973a: 68]. The transition here is most probably one of plenitude than transmutation, but by 1837, Geoffroy's views on evolution were widely known, as was Cuvier's demolition job on Lamarck.

LOUIS AGASSIZ: THE LAST FIXIST

Louis Agassiz, Cuvier's devotee and intellectual successor, concurred with Cuvier and Prichard on the constancy of species. Notwithstanding this, Agassiz did not expect that there would be a set of characters unique to all members of a species; resemblance was not itself clear or absolute. In an early short note [Agassiz 1842], he denied that characters gave the species and instead insisted that while there was a process that underlay the forms of species, there need not be any diagnosable characters that all members of the species exhibited, bringing to mind the Lockean distinction between real and nominal essences. Like Locke, Agassiz was rejecting the nominal and accepting the real essence, and it was a generative notion of real essence at that: *"[N]o so-termed character—that is, no observable mark—can be so striking as to indicate an absolute specific distinction; but at the same time, it should never be regarded as so trifling as to point to absolute identity; that characters do not mark off species, but that the combined relations to the external world in all circumstances of life do"* [italics in original].

Agassiz's version of the real essence here is the causal relations of the organism throughout its life cycle to its environment, and he cites the occasional inclusion of the male and female of a species in separate taxa. Agassiz explicitly rejected a diagnostic notion of species and in effect anticipated the later notion of "cryptic species" of Mayr. However, in practice, he was not so exact. He distinguished between eight "species" of man—Caucasian, Arctic, Mongol, American Indian, Negro, Hottentot, Malayan, and Australian—and claimed that these were all independent creations, not related by descent, each with its own region, flora, and fauna. The basis for this was not some generative notion, or reproductive isolation (since it was clear that human "types" could interbreed without trouble), but their clear physical differences [Lurie 1960: 264ff.]. In short, the characters indicated an absolute specific distinction. Or perhaps man was different.

Agassiz was not a Lockean, however; he was clearly a variety of Platonist, or at least of idealist. He wrote, "There is a system in nature . . . to which the different systems of authors are successive approximations. . . . This growing co-incidence between our systems and that of nature shows . . . the identity of the operations of the human and the Divine intellect" [Agassiz 1859: 31; Lurie 1960].

Agassiz's biographer Lurie calls this Agassiz's "cosmic philosophy" and notes that in his view:

[s]pecies, the individual units of identity in nature, were types of thought reflecting an ideal, immaterial inspiration. The same was true of the larger taxonomic categories—genera, families, orders, branches, and kingdoms. All such categories had no real existence in nature. Reality could be discovered only in the character of the individual animals and plants that

had inhabited or were now inhabiting the material world. The individual fossil or living from represented on earth the categories of divine thought ranging from species to kingdom and ultimately symbolized a complete identity with the highest concept of being, God. [Lurie 1960]

Two years after the publication of his "Essay on Classification," the *Origin of Species* was released, and so the essay provides a good demarcation point between the traditional view of classification, and the revolution that was to come, even if Agassiz's views were already archaic. In that essay,[9] he argued again for the stability of species, although his primary task was to discuss ways in which naturalists could identify and name species, rather than to define them other than as the smallest division of the four great *embranchements* named by Cuvier, which Agassiz called "great types" [Winsor 1979: 97]. These were the ways of being, typical plans of nature. Species were the lowest group that could be differentiated out of these plans (which play the role, therefore of Aristotle's summa genera). They had in themselves no identifying morphological character, because that was exhausted in the genus. Winsor notes [1979: 98] that

> [h]aving already publicly rejected the criterion of interbreeding, during the debate on the unit of mankind, Agassiz had to ask himself what besides morphological detail and sexual preference enables a biological species to be recognized. His answer was, its mode of reproduction and growth, its geographic distribution and fossil history, and the manifold relations that the individual organism bears to the world around it.

A species is, it seems, a description of the overall biological features of organisms, for Agassiz gave the individual organism priority, not unlike Buffon. Winsor notes further, "[h]is purpose was . . . to affirm the reality of all those relationships of similarity that are expressed in a natural classification." She quotes him from the essay:

> Species then exist in nature in the same manner as any other groups, they are quite as ideal in the mode of existence as genera, families, etc., or quite as real. . . . Now as truly as individuals, while they exist, represent their species for the time being and do not constitute them, so truly do these same individuals represent at the same time their genus, their family, their order, their class, and their type, the characters of which they bear as indelibly as those of the species.

Species exist as ideas, which represent the relations actual individuals bear to the world. They are not things, in the physical sense of the term, so much as what the things represent (but do not comprise). This is very Platonic.

In his *Methods of Study in Natural History* [Agassiz 1863], he further discusses classificatory categories. In this he adopts a realist view of Cuvier's *embranchements* theory as "being, so far as it is accurate, the literal interpreter of [the plan of creation]" [p. 41], and "that classification, rightly understood, means simply the creative plan of God as expressed in organic terms" [p. 42], but this is at a much higher level than the species level. He notes in chapter 5 that the nature of Linnaean Orders is partly arbitrary, that if "one man holds a certain kind of structural characters superior to another, he will establish the rank of the order upon that feature" [p. 72f.]; but overall, he says that higher taxa "stand, as an average, relatively to each other, lower and higher" [p. 86; cf. also p. 109 on branches and classes, and p. 127f. on genera]; he is a rank realist. He argues that there is no succession between these higher types [p. 92], and so there is no reason to think evolution is correct.

In this post-*Origin* work, Agassiz denies that species are more real than the higher taxa, contrary to the opinions based on de Candolle's views [p. 135]. Species are exactly as real as these higher taxa. "All the more comprehensive groups," he says, "equally with Species, are based upon a positive, permanent, specific principle, maintained generation after generation with all its essential characteristics. Individuals are the transient representatives of all these organic principles, which certainly have an independent, immaterial existence, since they outlive the individuals that embody them, and are no less real after the generation that has represented them for a time has passed away, than they were before" [p. 136]. Species are not composed of organisms, in other words; organisms at best "represent" species. And species are of the same standing as the higher taxa, which are "built upon a precise and definite plan which characterizes its Branch,—that that plan is executed in each individual in a particular way which characterizes its Class," and so on, down to species [139f.].

He discusses variation in domesticated animals ("which has been urged with great persistency in recent discussions" [i.e., by Darwin and his followers; p. 141] but asserts that this is due to the "fostering care" of the breeders of freaks that are not observed in wild species [p. 145] and that "this in no way alters the character of the Species" [p. 141F.]. "They are called Breeds, and Breeds among animals are the work of man: Species were created by God" [p. 147].[10] Homologies, the foundation of Darwinian argument, are "the Creative Ideas in living reality" [p. 231].

All is the work of God, and we are just making classifications to trace what God hath wrought.

Later, Agassiz further attacked "Darwinism" by means of an attack on Haeckel's genealogical classifications in Darwin's name, in a chapter added to a French edition [Agassiz 1869; Morris 1997]. Here he attacks the a priorism of the work of Oken and those who follow the ideal morphology school, including Haeckel in that class because he imposes his expectations on the data, which he also accuses Darwin of doing, a point he had made in an earlier review of the *Origin*. He rejects the claim of the "Darwinists and their henchmen" that organisms will not reproduce the essential characters of their ancestors. In Morris's translation,[11] Agassiz says. "All the observations relative to domestic animals, among which there are so many and so numerous variations, again did not succeed in demonstrating a sufficiently large amplitude in these variations; never did they [the Darwinists] have as a result anything which manifests the indefinite tendency to a changeability without limit."

In short, Agassiz rejected the Darwinian view of species on the grounds that the requisite variation had not been observed. Oddly, Agassiz was unable to see that there could possibly be more than a certain amount of variation in his own specimens, as Lurie [1960: 194f.] noted:

> When he had hundreds of fishes spread before him on a work table, these convictions (of the fixity of species) were of such force that even his keen powers of observation and his excellent ability to compare diverse types failed him. He insisted on identifying specimens that seemed even the slightest degree different from one another as separate species rather than as variants. In one analysis alone, for example, he described nine separate "species" of fishes that were in actual fact reducible to four schools of single species.[12]

Agassiz, merciless on taxonomic splitters, was in practice himself a splitter because of his tendency to classify on form alone despite his stated convictions about species in theory. In another, more sinister respect, his views also led him to excessive splitting. He claimed that Negroes were not of the same species as whites, because, it seems, of the feelings of revulsion he had for them that led him to deny they could be conspecific to whites. This famously meant that he became the leading proponent of multigenism and hence a popular figure in the South before the Civil War [Hunter Dupree 1968: 228f.].

Amundson [2005: 79] argues that Mayr's access to Agassiz's works at the Museum of Comparative Zoology at Harvard, which Agassiz had founded and of which Mayr was a later director, led Mayr to overgeneralize that all taxonomists before Darwin were essentialists, typologists, and fixists. Agassiz was indeed a Platonist who considered species thoughts in the mind of God, and he was also undoubtedly a species fixist, particularly after Darwin. Moreover, he was both a taxonomic essentialist and a material essentialist. Possibly he was the *first* such essentialist fixist Platonist. But he was not, so to speak, typical of the time or the profession, apart from his students, and they only for a limited time.

JAMES DANA: A LAW OF CREATION

But Agassiz was not alone in pressing the Cuvierian view in the period leading up to the *Origin*. His very great admirer James Dana, in an essay in a journal he coedited [Dana 1857], reiterated the old view that "species" applies to all natural things, and that the variable characters of individuals are merely confusing. To this end, he rejected the idea that species are even groups, necessarily. Instead, he wrote, "A *species* corresponds to *a specific amount or condition of concentrated force, defined in the act or law of creation*" [p. 306, italics in original].

At least in the inorganic world: species are what they were constituted at their creation to be. In the biological world, the idea is the same, leading to the understanding that

> [t]he species is not the adult resultant of growth, nor the initial germ cell, nor its condition at any other point; it comprises the whole history of development. Each species has its own special mode of development as well as ultimate form or result, its serial unfolding, inworking and outflowing; so that the precise nature of the potentiality in each is expressed by the line that historical progress from the germ to the full expansion of its powers, and the realization of the end of its being. We comprehend the type-idea only when we understand the cycle of evolution [sensu *development*—JSW] through all its laws of progress, both as regards the living structure under development within, and its successive relations to the external world. [p. 308]

For Dana, species are the units of the organic world as molecules are the units of the inorganic. He discusses the ranges of infertility of hybrids from the infertile mule to the continuously fertile hybrid and says that the fully fertile hybrid is not observed in nature, at least among animals; plants are more frequently hybridizing. In a rather backhanded manner, he affirms the monogenist position—humans are one species, although nonwhite races are disappearing "like plants beneath those of stronger root and growth, being depressed morally, intellectually and physically, contaminated by new vices, tainted variously by foreign disease, and dwindled in all their hopes and aims and means of progress, through an overshadowing race" [p. 311]—lest any of his readers get the wrong idea. At least he stood up to Agassiz on monogenism.

Species are not transmutable, for all hybridization is merely recombination of already extant variation, but there is variation within species—the unfolding of the potentiality inherent within a species according to natural law and changing circumstances [p. 312]. Species are liable to variation as part of the law of a species, and knowledge of the complete type requires knowing all these and how they relate to external circumstances. There is a higher essence, as it were, in the type. Finally, while species are real things, they are not comprehensively covered in any "material or immaterial existence"—in modern parlance, they are types, not tokens,[13] and species are both invariant and variant. In short, Dana sees species as the schematic of a developmental cycle and the ways in which it may be perturbed by the environment.

RICHARD OWEN: THE UNITY OF TYPES

Richard Owen, who introduced many of the ideas of the ideal morphologists into British thought, did not address the question of species directly, so far as I can tell. In his Hunter Lectures of 1843, reissued in a revised edition in 1855 entitled *Lectures in Comparative Anatomy*, Owen addresses the question of overall types in terms derived from Cuvier's *embranchments* but treats species as unanalyzed units of classification. He was not opposed at this time to species transmutation, although he treats it more as a formal possibility than as an actuality, and so he contributes little to the topic in hand.

Owen, as is widely known, first clearly expressed the distinction between *homologue* and *analogue* in the glossary to his 1843 *Lectures on Invertebrate Animals* and further distinguished special, general, and serial homology—respectively, the correspondence of a part in one animal with a part in another, of a part in a particular animal (that is, of a species) to a fundamental or general higher type, and the repetition of a part within a particular animal or type [Russell 1982: 108f.]. Organic forms, he thought, were due to a mutual antagonism of two principles, one of which brought about a vegetative repetition of structure, and the other that shapes the living thing to its function, a teleological principle [Russell 1982: 111; cf. Amundson 2005: 88–93]. He allowed that species were formed through time and that there was a naturalistic cause based on these principles, "the secret counsels of the organizing forces," as he expressed it in the *Archetype* [Owen 1848; see the discussion in

Amundson 2005, chapter 4]. Earlier, Owen discussed how species have changed in their meaning since the times in which Linnaeus's and Cuvier's definitions were accepted [Owen 1835]:

> I apprehend that few naturalists now-a-days, in describing and proposing a name for what they call "a new *species*," use that term to signify what was meant by it twenty or thirty years ago, that is, an originally distinct creation, maintaining its primitive distinction by obstructive generative peculiarities. The proposer of the new species now intends to state no more than he actually knows; as for example, that the differences in which he founds the specific character are constant in individuals of both sexes, so far as observation has reached; and that they are not due to domestication or to artificially superinduced external circumstances, or to any outward influence within his cognizance; that the species is wild, or is such as it appears by nature. [quoted in Huxley 1906: 303]

Owen here treats species as taxonomic objects, with the underlying implication that they are constant due to natural causal powers. The generative "peculiarities" are not ignored or forgotten when the creation part of the definition has been abandoned, and the naturalizing of species is evident. However, it is also noteworthy given Owen's later reputation that in his presidential address to the British Association for the Advancement of Science in 1858, Owen not only reports Darwin's and Wallace's views on speciation by natural selection but goes on to say:

> No doubt the type-form of any species is that which is best adapted to the conditions under which such species at the time exists; and so long as those conditions remain unchanged, so long will the type remain; all varieties departing therefrom being in the same ratio less adapted to the evironing conditions of existence. But, if those conditions change, then the variety of the species at an antecedent date and state of things will become the type-form of the species at a later date, and in an altered state of things. [Basalla, Coleman, and Kargon 1970: 329]

Owen's initial reaction to Darwin and Wallace, untainted by Huxley's rhetoric, was thus fairly positive. Of special note is that Owen refers to the "type-form" of the species in a way that makes it clear that he is not referring to essences.

Ronald Amundson [2005: 81] has argued that for the period, species were never types, as types were something that united species under larger classificatory ranks. Indeed, Amundson denies that species ever were types, a point Owen, among others, shows is not correct. As many have noted, there is a distinction between Platonic forms and Aristotelian species, and the Aristotelian form *could* be varied from the standard type.

Naturalists of a Platonic bent prior to Darwin, such as Agassiz, saw species as types from which individuals could deviate. The concept of "degeneration," which Amundson discusses [pp. 36, 40], indicates that individuals were able to improperly instantiate the specific norm, as well as species as units being able to improperly or variably instantiate the generic essence. So long as types are not identified with unvarying essences, it is not true that species could not be types. But if, as Amundson seems to do in his section 4.2, types (here, morphological types as abstractions) and essences are identified, with Mayr, then there are no specific types.

That said, it is apparent both that types are almost always through this period treated as large-scale (supraspecific) types and that the ideal morphologists in the tradition of Goethe and Oken did not generally discuss species, either as biological realities or as an abstract concept—they were more concerned with the higher types that gave a Unity of Type between species and genera [Russell 1982; Nyhart 1995].

OTHER FIXIST VIEWS

The view that species were fixed as they had been created was held religiously (in the strict sense) in 1844 by the marine biologist and founder of the fashion of aquarium keeping, Philip Henry Gosse in his *An Introduction to Zoology* [p. xv, cited in Simpson 1925: 175]:

> Each order was distributed into subordinate groups, called Genera, and each genus into Species. As this last term is often somewhat vaguely used, it may not be useless to define its acceptation. It is used to signify those distinct forms which are believed to have proceeded direct from the creating hand of God, and on which was impressed a certain individuality, destined to pass down through all succeeding generations, without loss and without confusion. Thus the Horse and the Ass, the Tiger and the Leopard, the Goose and the Duck, though closely allied in form, are believed to have descended from no common parentage, however remote, but to have been primary forms of the original creation. It is often difficult in practice to determine the difference or identity of species; as we know of no fixed principle on which to found our decision, except the great law of nature, by which specific individuality is preserved—that the progeny of mixed species shall not be fertile inter se.

Gosse repeated this in his famous book [Gosse 1857], known popularly as *Omphalos* (also known as *Creation [Omphalos]*), in which he argued that the world was created as the Bible said but was made to look old. He says:

I demand also [as well as the creation of matter out of nothing—JSW], in opposition to the development hypothesis [pre-Darwinian evolutionism—JSW], the perpetuity of specific characters, from the moment when the respective creatures were called into being, till they cease to be. I assume that each organism which the Creator educed was stamped with an indelible specific character, which made it what it was, and distinguished it from everything else, however near or like. I assume that such character has been, and is, indelible and immutable; that the characters which distinguish species from species *now*, were as definite at the first instant of their creation as now, and are as distinct now as they were then. [p. 111][14]

Here we see both fixism and essentialism, and it is essentialism of the kind that Mayr and the Received View objected to—a *real* or material essentialism as well as a taxonomic one. But there is no comfort for the Received View here—Gosse was not regarded as a leading professional naturalist [though Huxley called him the "honest hodman of science"; Numbers 1992: 141], nor was his book well received; it sold so poorly, most copies were pulped, and as his son wrote, "alas! Atheists and Christians alike looked at it, and laughed, and threw it away" [Gosse 1970: 68]. It is hard not to see Gosse's obduracy as a desperate attempt to maintain the fixity of species even in the face of Buffon's hybridization experiments, which had had a partial success, but in fact there was considerable blindness to variation, as the Agassiz example shows. Another instance cited by J. Y. Simpson [p. 178f.] involves a student of Agassiz, Stimpson, who, when finding intermediate forms of a mollusk he could not decide to place in one species or another, "after he had studied it for a long time, put his heel upon it and grind[ing] it to powder, remarking, 'That's the proper way to serve a damned transitional form'" [quoting Nathaniel Southgate Shaler, another student of Agassiz who ended up less disposed to essentialist accounts, from his *Autobiography*, p. 129].

CHARLES LYELL: SPECIES ARE FIXED AND REAL

More significantly, for our later story, are the earlier views of Charles Lyell. It is well known that it was the second volume of Lyell's *Principles of Geology* [Lyell 1832] that Darwin received on the voyage of the HMS *Beagle* [Kottler 1978: 276–278], which contained the discussion of Lamarck's views, a "book-long refutation of Lamarck," as Desmond and Moore called it [1991: 131]. Lyell presented the now standard fixist view of species; as Kottler put it, "Lamarck and Lyell agreed that the 'reality' of species implied their constancy. The words 'real' and 'permanent' were synonymous with respect to species in nature. Thus while Lamarck, the transmutationist, contended species were not real, Lyell, the fixist, argued they were" [p. 277]. In this volume, Lyell concludes his discussion of Lamarck and Linnaeus and species in general, at the end of chapter 4 with this:

> For the reasons, therefore, detailed in this and the two preceding chapters, we draw the following inferences, in regard to the reality of species in nature.
>
> First, That there is a capacity in all species to accommodate themselves, to a certain extent, to a change of external circumstances, this extent varying greatly according to the species.
>
> 2dly. When the change of situation which they can endure is great, it is usually attended by some modifications of the form, colour, size, structure, or other particulars; but the mutations thus superinduced are governed by constant laws, and the capability of so varying forms part of the permanent specific character.
>
> 3dly. Some acquired peculiarities of form, structure, and instinct, are transmissible to the offspring; but these consist of such qualities and attributes only as are intimately related to the natural wants and propensities of the species.
>
> 4thly. The entire variation from the original type, which any given kind of change can produce, may usually be effected in a brief period of time, after which no farther deviation can be obtained by continuing to alter the circumstances, though ever so gradually,—indefinite divergence, either in the way of improvement or deterioration, being prevented, and the least possible excess beyond the defined limits being fatal to the existence of the individual.
>
> 5thly. The intermixture of distinct species is guarded against by the aversion of the individuals composing them to sexual union, or by the sterility of the mule offspring. It does not appear that true hybrid races have ever been perpetuated for several generations, even by the assistance of man; for the cases usually cited relate to the crossing of mules with individuals of pure species, and not to the intermixture of hybrid with hybrid.
>
> 6thly. From the above considerations, it appears that species have a real existence in nature, and that each was endowed, at the time of its creation, with the attributes and organization by which it is now distinguished.
> [p. 64f. in the 1832 edition]

Kottler also considers the fifth edition of the *Principles*, published in 1837, which upon his return and as he began to consider transmutation and hence the nature of species, Darwin heavily underlined and annotated. As before, Lyell relies on the infertility of hybrids. According to him, no hybrid could give rise to a new species, unless it was back-bred into a pure species. In nature, species had an aversion to interbreeding, which made it "a good test of the distinctness of original stocks, or *species*" [quoted in Kottler 1978: 277 from Volume I, p. 523].

While he (reluctantly) changed his mind after the publication of the *Origin* and many discussions with Darwin directly, his major contribution at this time is to affirm the fixity and reality of species. In dealing with the views of Lamarck and Geoffroy over transmutation, he noted:

> The name of species, observes Lamarck, has been usually applied to every collection of similar individuals, produced by other individuals like themselves. This definition, he admits, is correct, because every living individual bears a very close resemblance to those from which it springs. But this is not all which is usually implied by the term species, for the majority of naturalists agree with Linnaeus in supposing that all the individuals propagated from one stock have certain distinguishing characters in common which will never vary, and which have remained the same since the creation of each species. [p. 2]

So, Lamarck must defeat this by finding no gaps as we advance in our knowledge, and so our taxonomic characters are arbitrary. Lyell spends considerable time dealing with Lamarck's claims, resulting in the statement that the Author of Nature would foresee all conditions in which a species would exist, and so the changes Lamarck requires will not occur [p. 23f.]. Mayr [1982: 405] quotes him as saying:

> There are fixed limits beyond which the descendants from common parents can never deviate from a common type. . . . It is idle . . . to dispute about the abstract possibility of the conversion of one species into another, when there are known causes, so much more active in their nature, which must always intervene and prevent the actual accomplishment of such conversions. [From the 1835, 3rd ed., Volume II, p. 162; the second sentence is not found in the 1st ed.]

Mayr thinks this is an expression of Lyell's essentialism: "each species had its own specific essence and thus it was impossible that it could change or evolve. This, for example, was the cornerstone of Lyell's thought" [Mayr 1982]. But was it really? Lyell seems to be saying not that an *essence* is causing it to remain stable, but that a species is held stable by interbreeding and "known causes" of infertility. There is typology, to be sure, but overall, Lyell's view is a causal one: again, we see here hints of a generative notion of species. However, Mayr correctly notes that the *Principles* was Darwin's scientific "bible" [p. 406] and that he devoted so much time in the *Origin* to refuting special creation largely because of the challenges set by Lyell [p. 407].

A.-P. DE CANDOLLE AND ASA GRAY: THE BOTANICAL VIEW

A third stream of thought in this period supposes that species are real and that so also is variation from the type. This is primarily due to the famous family of Swiss botanists, the de Candolles, in particular the elder, Augustin-Pyramus [Candolle 1819]. De Candolle stressed the variation of living things, and defined species as "the collection of all the individuals who resemble one another more than they resemble others; who are able, by reciprocal fecundation, to produce fertile individuals; and who reproduce by generation, such kind as one may by analogy suppose that all came down originally from one single individual" [quoted in Hunter Dupree 1968: 54].

So, for de Candolle, species are groups of individual organisms. They are, in the tradition of Ray, both to resemble one another and to generate progeny that are fertile and resemble one another. As did Buffon and the older tradition, de Candolle treated variation as the effect of local environments and occasional hybridization. In this opinion, he was followed closely by the great American botanist, Asa Gray, who was later to become significant in the promotion of Darwinian theory in America against Agassiz. As late as 1846, in a review of the *Vestiges of Natural Creation*, Gray declared that species were created as they are found, and did not transmute [Hunter Dupree 1968: 145–147]. But by the 1850s, he had an operational view of species. While creation may have once been of importance, writes his biographer Hunter Dupree [1968: 217], what most concerned Gray was that if species transmuted as Lamarck, Geoffroy, and the author of the *Vestiges* declared, then natural history would become meaningless, one presumes because we would be unable to specify the facts about the groups in biology that we encounter. Hunter Dupree says that "it was this inability [of unlike species to breed together—JSW] which created the species border, not that he or any other could find this border easily, least of all by referring to an ideal type." Morphology was only a guide to these borders, and relied on the experience of the naturalist and the principles of classification. The Lockean character of this account is manifest. The real essence here is interfertility, not morphology. As a result, Gray worried about hybridization and its role in speciation. This had been a concern since Linnaeus's time, and the urgency of the problem was progressively increasing among the naturalists of the period. Gray noted that hybrids would stand a good chance of being fertilized by their parents and asked, "In such cases they are said to revert to the type of the species of the impregnating parent; but would they return exactly to that type, inheriting as they do a portion of the blood of a cognate species?" The modern problem of the introgression of genes into a species is foreshadowed here, although Gray relies on a blending inheritance model, of course, causing swamping of the variations.

Gray's other contribution to this topic is in his assertion that humans are a single species. He felt that science in general and his ideas on species and hybridization in particular pointed to the unity of the human race [p. 220]. It is worth noting here that Gray's rather orthodox Protestantism seems to have had no particular impact on his view of species, as Cuvier's also had not. Objections to transmutation appear not to have been founded on orthodox religious doctrine.

PRE-DARWINIAN EVOLUTIONARY VIEWS OF SPECIES

> [S]pecies, the subdivision where intermarriage or
> breeding is usually considered as natural to animals,
> and where a resemblance of offspring to parents is
> generally persevered in.
> **Vestiges of the History of Creation [Chambers
> 1844: 263]**

In German- and French-speaking countries prior to the publication of the *Origin*, there was a number of specialists propounding evolutionary views of species. Of note are Bonaparte and Unger.

Prince Charles Lucien Bonaparte (1803–1857), a nephew of the famous Napoleon, was an active ornithologist, as his father Lucien had been after the British released him from detention in 1814 [Stresemann 1975]. Exiled in Leiden by his cousin Louis Napoleon, he became friends with Hermann Schlegel, another famous ornithologist. In 1851, he published the first volume of his *Conspectus generum avium*, a survey of all known species of birds worldwide. In this work, he treated extant species as the descendant forms of prior extinct forms, and in an address to an 1856 convention on "What Is a Species, Particularly in Ornithology?" he said:

> We will state with unanimous conviction that the antediluvian crocodiles, elephants and rhinoceroses were the ancestors of those living in our day, and these animals would not have been able to continue to exist without the manifold mutations that their systems produced to adapt themselves to the environment, and that became second nature to their descendants. . . . If the environment remains the same, so do the species. The stabilizing influence is then by itself all-powerful. The mutating influence can succeed in opposing it only when the whole world surrounding it changes. . . . But

races, however different in characteristics they may be, vanish entirely or at least do not long survive as soon as the environment that produced them ceases to be the same. . . . The transitions between the different races and their type are the best evidence that we can supply to set aside putative species, which are to be relegated to races, with which the painstaking zoologist must nevertheless occupy himself just as earnestly. [quoted in Stresemann 1975: 166]

Of interest in this excerpt is the implication that species are racial groups stabilized by the influence of the environment, somewhat as stabilizing selection operates (although there is no reason to suppose Bonaparte thought selection was the reason for the stabilization). As Darwin later also argued, races are merely species in the making that are not yet made stable. Bonaparte died in 1857, leaving the *Conspectus* unfinished.

Franz Unger, an Austrian botanist at the University of Vienna, published a form of common descent with modification theory in 1852, entitled "Attempt at a History of the Vegetable Realm" *(Versuch einer Geschichte der Planzenwelt)*, in which he supposed that all plant life was a single entity that had developed new forms. Unger's theory is often taken to be a forerunner of Darwin [cf. Temkin 1959: 339–342], but he in fact thought that all subsequent development was an expression of the original potentiality of the *Urpflanze*: "Nothing has been added in this regulated evolutionary process of the vegetal world that had not been previously prepared and indicated, so to speak. Neither genus, nor family, nor class of plants has manifested itself without having become necessary in time" [quoted in Temkin 1959: 340].

Temkin notes that Unger held that species themselves do not change but that some individuals metamorphosed while the old type remained in existence for some time. One year later, Hermann Schaffenhausen published an article "On the Constancy and Transformation of Species" rebutting Unger's ideas on the grounds they indicated man evolved from an orangutan. However, he said in his summary [quoted in Temkin 1959: 342] that "[t]he immutability of species which most scientists regard as natural law is not proved, for there are no definite and unchangeable characteristics of the species, and the borderline between species and subspecies [*Art* and *Abart*] is wavering and uncertain."

It appears, then, that in the post-Romantic period in Germany and German-speaking countries, naturalists were not so rigid over species as was the Swiss export to America, Agassiz. Unger's conception appears to be an entelechical view—a species was a type that was "in" the plant kingdom from the beginning, in the *Urpflanze*. However, the stasis of the species themselves is due in Unger's book to the generative powers of inner forces. Mayr [1982: 391] quotes him as saying, "The lower as well as the higher taxa appear then not as an accidental aggregate, as an arbitrary mental construct but united with each other in a genetic manner and thus form a true intrinsic unit."

Finally, mention must be made of Heinrich Georg Bronn (1800–1862), who, in his prize-winning submission to the Paris Academy of Sciences in 1857 published the *Untersuchungen über die Entwicklungs-Gesetze der organischen Welt während der Bildungs-Zeit unserer Erd-Oberfläche* (Researches into the Laws of Development of the Organic World during the Period of Development of Our Earth's Surface), in which he presented a tree diagram for the progressive evolutionary divergence of types [Nyhart 1995: 110–116; Panchen 1992: 26f.]. Bronn, Nyhart tells us, was committed to the progressionism of Oken and Lamarck, and when he supervised the translation of Darwin's *Origin*, he translated "favoured races" in the subtitle as *"vervollkommneten Racen,"* or "perfect[ed] races," a subtlety that may have influence Haeckel's later view of evolution [Junker 1991]. Bronn had previously, in his 1841 *Handbuch einer Geschichte der Natur*, treated species as the result of acts of special creation.

However, he appended a critical essay to chapter 15 in which he said that he doubted varieties would permanently branch off, and instead he held that species were not transformed through inherited modifications but by a law of nature, a creative force, as yet unknown [Nyhart 1995: 112f.]. Each species had its own life span, and then a more perfect one replaced it.

JOSEPH HOOKER, THOMAS WOLLASTON, AND GEORGE BENTHAM: LOGIC AND DIVISION

Two important writers on species, Joseph Hooker and George Bentham, were collaborators, but each had a different approach. Hooker was one of Darwin's closest confidants.[15] He was introduced to Darwin's views on evolution as early as the 1844 manuscript, yet he was unable to discuss these ideas with anyone but Darwin. When he finally was able to, after the 1858 reading of Wallace's and Darwin's papers, he noted to Asa Gray that he could never "allude to his doctrine in public, & I always had in my writings to discuss the subject of variation etc & as if I had never heard of Natural Selection—which I have all along known & feel not only useful in itself as explaining many facts in variation, but as the most fatal argument about 'Special Creation'" [quoted in Stevens 1997: 346].

Hooker considered that much variation in plants was intraspecific and that there were far fewer actual species than were listed and that what he called its "habit"—or general appearance and growth—was deceptive as a guide to the difference between species and varieties [Stevens 1997]. He wrote to Gray in 1856, "As to consistency with regard to species—it is a myth, a delusion . . . the most consistent men are hair-splitters—they make almost every difference specific." In the light of Whately and the logic of division in which most had been trained, at least implicitly, this tension is understandable. Every difference *was* specific, logically. The problem lay in that in natural history, specifically in botany, the line was drawn much higher than that: "Bother variation, developement [sic] & all such subjects,! It is reasoning in a circle after all. As a Botanist I must be content to take species as *they appear to be* not as *they are,* & still less as they were or ought to be" [to Darwin, July 1845; quoted in Stevens 1997: 349].

Thomas Vernon Wollaston, in his 1856 *On the Variation of Species* [England 1997], defined species as a community of descent within which there was variation but between which there was no gradation. Those who Hooker had called hair-splitters he called "very hyper-accurate definers." Nevertheless, after Darwin published the *Origin,* Wollaston denied the indefinite variation that Darwin needed existed. Hooker, to the contrary, was convinced it did. In 1853, in the introduction to his *Botany of the Antarctic Voyage,* volume 2, he noted that we needed to "act upon the idea that for practical purposes at any rate species are constant" [Stevens 1997] but observed that variation tended to be absent in a particular place or colony but could often be found globally. He noted, in a passage that Gray marginally wrote was "opt[issime]!" in his copy [Stevens 1997: 364, n. 62]:

> It is very much to be wished that the local botanist should commence his studies upon a diametrically opposite principle to that upon which he now proceeds, and that he should endeavour, by selecting good suites of specimens, to determine *how few,* not *how many* species are comprised in the flora of his district. The permanent differences will, he may depend upon it, soon force themselves upon his attention, whilst those which are non-essential will consecutively be eliminated. There is no better way of proving the validity of characters than by attempting to invalidate them.

The professional botanist's role here was crucial. Hooker, George Bentham, Ferdinand Mueller, and others were better placed by virtue of having specimens from across districts and indeed the globe to mark out species, than local observers or gardeners. Bentham, whose work in logic was both traditional (dichotomous) and radical (he introduced the idea of quantification in his 1827 *Outline of a New System of Logic* [Bentham 1827; McOuat 2003], written in reply to Whately's *Elements*), early on had a fairly traditional view of species in natural history—they were created by God, of course, but he took more notice of the Cuvierian definition of descent from a common ancestor. In his early work as a botanist, a field he turned to after his logic book failed to garner much contemporary attention (it got rather more later, when the issue of who invented logical quantification was being discussed), he was convinced that species could be ranked by a comprehensive survey of variation [Stevens 1997: 359]. After Darwin, he was not so sure, and eventually he gave up the idea of a species rank. In 1864, he wrote to Ferdinand von Mueller that "true species are entirely limited in nature." As Stevens notes, it made little or no difference to his taxonomic work. However, in an anonymous review of de Candolle's *Geographie botanique raisonée* in 1856, he had stated clearly enough that a species was "a collection of individuals which, by their resemblance to each other, or by other circumstances, we are induced to believe are descended or *may have* descended from one individual or a pair of individuals" [quoted in Stevens 1997: 360]. What the inducements were, however, were unclear. Bentham almost indicates that *species* is a purely operational notion.

A SUMMARY VIEW OF THE EARLY NINETEENTH CENTURY

It appears that while many naturalists were fixists, the leading criterion for species identification or explanation was derived from the descent of similar forms. Apart from Agassiz, nobody seems, however, to have inferred from fixism, or the pious creationism that was the usual form of words used, that species had essences or even that variation was firmly limited. In this period, variation was a real research difficulty. Cuvier's definition was widely disseminated and repeated, and Linnaeus's almost taken as something too obvious, and a little overreligious, to mention. This background is something shared by Darwin and serves to highlight both his orthodoxy in this regard and those areas in which he innovated.

DARWIN AND THE DARWINIANS

One of the ironies of the history of biology is that Darwin
did not really explain the origin of new species in *The
Origin of Species*, because he didn't know how to define
a species.
Douglas J. Futuyma [1983: 152]

The Origin of Species, whose title and first paragraph
imply that Darwin will have much to say about speciation.
Yet his magnum opus remains largely silent on the
"mystery of mysteries," and the little it does say about this
mystery is seen by most modern evolutionists as muddled
or wrong.
Jerry A. Coyne and Allen H. Orr [2004: 9]

Darwin's ideas have been widely misinterpreted almost from the date of
the publication of the *Origin* in November 1859. In this chapter, we shall
see that he has in fact a fairly orthodox view of species as real things in
nature (albeit temporary things), that he did not think interfertility was
a good test of a species, and that his dismissive comments in the *Origin*
have more to do with the professional nature of taxonomy and the difficulties of diagnosis and nomenclature than a claim that species did not
exist at all.

• • •

It is occasionally stated that Darwin denied the reality of species or held
that "species" was an arbitrary concept (see the epigraphs opening this chapter). Mayr notes "one might get the impression [from the *Origin of Species*]

that he considered species as something purely arbitrary and invented merely for the convenience of taxonomists" [Mayr 1982: 268]. Mayr goes on to note that he nevertheless treated species in a perfectly orthodox taxonomic manner and that he treated the concept purely typologically. Beatty [1985] and Ereshefshy [1999] raise similar doubts on Darwin's view of species. On the contrary, it will be argued in this chapter that Darwin was a species realist, and although he developed his views over time, he never ceased being a species realist, just as his mentor Lyell was.

Charles Darwin is important not so much for the novelties on the nature of the species concept that he provided—there are only really two of these, failure to breed in nature, and selection as the motive force of specific characters, as we shall see. Rather, it is because his book *On the Origin of Species* changed *every* scientist's way of looking at species thereafter. He has been more closely scrutinized than anybody else, and there is a wealth of material available. One thing that we should put out of our minds from the beginning, though: it is *not* true that Darwin did not address the origin of species in *On the Origin of Species*. The book is "one long argument" [Mayr 1991] on that very point. Over and again, he discusses why species evolve to be distinct from parental forms, and how they have done so. It is unclear to me how this idea gained currency.[1]

Darwin's views on species changed over time. In his earlier works he seems to have treated species the same way as his teachers, unsurprisingly, as groups united by some description—that which Mayr refers to as "typological, 'non-dimensional' species of local fauna" [Mayr 1982: 265]. Since his views are often misrepresented [see Kottler 1978 for a discussion, particularly p. 291f.] on the basis of comments made in the *Origin*, the following chapter will give an extensive and chronological series of quotations from his published works, including his correspondence and the Notebooks. While Kottler's analysis is reliable, some features of Darwin's views that are further developments of the older notions of *species* and some of his comments that are relevant to later debates, particularly over speciation, are to be found in comments not discussed by Kottler, Mayr, or Ghiselin [1984].

THE NOTEBOOKS

Darwin first began thinking about the nature of species in his Notebooks in 1837 and 1838. Kottler [1978] has investigated Darwin's early views on species in the Notebooks B, C, D, and E (labeled by Kottler I–IV, which I have relabeled conventionally). These are referred to as the "transmutation notebooks" since Darwin started them after he became convinced by Gould's investigations of the finches found on the Galápagos Islands, and the tortoises and mockingbirds backed it up, that these were modified descendents of South American colonists [Desmond and Moore 1991: 224f.]. Once he started on the idea, his conjectures came thick and fast: mammalian species were shorter lived than simpler forms because of their complexity; domestic animals were able to revert to the wild forms or at least live like them; perhaps species had a vital force and a fixed lifespan; and so on. In the Notebooks, he noted the "repugnance" of species to intercrossing (all quotations from Kottler):

> Repugnance generally to marriage before domestication, . . . marriage never probably excepting from strict domestication, offspring not fertile or at least most rarely and perhaps never fertile.—No offspring: physical impossibility to marriage. [B120]

> Instinctive feelings against other species for sexual ends. [B161]

> There is in nature a real repulsion amounting to impossibility holds good in plants between all different forms. [B189]

> The dislike of two species to each other is evidently an instinct; & this prevents breeding. [B197]

> The existence of wild close species of plants shows there is tendency to prevent the crossing of animals where there is much facility in crossing there comes the impediment of instinct. [E143f.]

Kottler observes that at this stage, Darwin agreed with Lyell that intercrossing was forced in domestication, and he made noninterbreeding a test of being a species:

> Now domestication depends on perversion of instincts . . . & therefore the one distinction of species would fail. [B197]

> Definition of species: one that remains at large with constant characters, together with beings of very near structure. [B213]

> My definition of species has nothing to do with hybridity, is simply, an instinctive impulse to keep separate, which no doubt be overcome, but until it is these animals are distinct species. [C161]

> A species as soon as once formed . . . , repugnance to intermarriage— settles it. [B24]

> Species formed . . . keep distinct, two species made. [B82]

Clearly, Darwin is more concerned with the behavior of organisms in natural conditions, not with the mere possibility of intercrossing. A species is to him at this stage an interbreeding group that is kept separate from other groups not only by the impossibility of hybridization, but also by the mating behaviors of each group. Hence, it is not a notion that can be lab tested—although Buffon had famously, and with some success, tested his idea that Linnaean-level species were geographic variants of the *première souche*, or primary stock (discussed earlier). It has to be observed in the field. But there were ways to test species: "It is daily happening, that naturalists describe animals as species. . . . There is only two ways [sic] of proving to them it is not; one where they can [be] proved descendant [Kottler interpolates: descent from common parents], which of course most rare, or when placed together they will breed" [B122].

The standard view of species at the time, since the original definition by Linnaeus, was that any two organisms were of the same species if they shared ancestry (in Linnaeus's pious formulation, from the pair of creatures created by God). Species are real, according to Darwin here, when they do not interbreed:

> As species is real thing with respect to contemporaries—fertility must settle it. [C152]

> If they [systematists] give up infertility in largest sense as test of species—they must deny species which is absurd. [E24].

Kottler notes that Darwin is not here using Buffon's 1749 definition of species, as by the phrase "in the largest sense," Darwin is including both sterility and aversion, which Buffon had not, merely requiring sterility when crossed. Immediately before the last passage, says Kottler, Darwin had written, "[O]ne species may have passed through a thousand changes, keep distinct from other, & if a first & last individual were put together, they would not according to all analogy breed together." Therefore, Darwin takes "being a species" as the *outcome of changes* that lead to a failure of the organisms to interbreed, not as the *outcome of failure to interbreed* first. He is not using a diagnostic notion of species. This is a generative notion, one to be explained by transmutation. In fact, Darwin expects that diagnosis may be nigh on impossible:

> Hence species may be good ones and differ scarcely in any external character. [B213]

> We do not know what amount of difference prevents breeding. [B241]

The Notebooks were completed eight years later, in 1845. In them Darwin developed a notion of speciation as due to adaptation to local conditions, mostly due to geographic isolation, which prevented backcrossing [Kottler 1978: 287f.].

DARWIN'S PRE-*ORIGIN* CORRESPONDENCE

In his correspondence with his scientific friends, Darwin begins to ask questions about species relatively late, around 1855 [Barlow 1967], although Padian [1999: 353] notes that in 1843 he did once ask the museum taxonomist G. R. Waterhouse what he meant by "relationship"; Darwin's query is instructive [July 26, 1843; Burkhardt 1996: 76]:

> It has long appeared to me, that the root of the difficulty in settling such questions as yours,—whether the number of species &c &c should enter as an element in settling the value of existence of a group—lies in our ignorance of what we are searching after in our natural classifications.—Linnaeus confesses profound ignorance.—Most authors say it is an endeavour to discover the laws according to which the Creator has willed to produce organized beings—But what empty high-sounding sentences these are—it does not mean order in time of creation, nor propinquity to any one type, as man.—in fact it means just nothing—According to my opinion, (which I give everyone leave to hoot at, like I should have, six years since, hooted at them, for holding like views) classification consists in grouping beings according to their actual *relationship*, ie, their consanguinuity, or descent from common stocks.

Waterhouse replied, "By relationship I mean merely resemblance." Darwin, having raised the issue of the number of species then treats species themselves as a subordinate issue—it is the ways in which *higher* taxa are to be arranged that he is most interested in. Shortly thereafter, he mentions species in passing to Hooker (January 11, 1844; Burkhardt 1996: 80):

> I was so struck with the distributions of Galapagos organisms &c &c & with the character of the American fossil mammifers, &c &c that I determined to collect blindly every sort of fact, which c^d bear any way on what are species. . . . At last gleams of light have come, & I am almost convinced (quite contrary to the opinion I started with) that species are not (it is like confessing a murder) immutable.

Some years later, he wrote to Hooker about the practical impact of the "question of species": "How painfully (to me) true is your remark, that no one has hardly a right to examine the question of species who has not minutely described many" [Darwin to Hooker, September 1849; Darwin 1888: 39]. Yet, in 1853, only four years afterward, Darwin noted to Hooker that his ideas on species had not made all that much difference to his classificatory work:

> [I]n my own work, I have not felt conscious that disbelieving in the *permanence* of species has made much difference one way or the other; in

some few cases (if publishing avowedly on doctrine on non-permanence) I sh^d. *not* have affixed names, & in some few cases sh^d. have affixed names to remarkable varieties. Certainly I have felt it humiliating, discussing & doubting & examining over & over again, when in my own mind, the only doubt has been, whether the forms varied *today or yesterday* (to put a fine point on it, as Snagsby would say). After describing a set of forms, as distinct species, tearing up my M.S., & then making them one again (which has happened to me) I have gnashed my teeth, cursed species, & asked what sin I had committed to be so punished: But I must confess, that perhaps the same thing w^d. have happened to me on any scheme of work—. [Darwin to Hooker, September 25, 1853; Burkhardt 1996: 128–129]

He asked Henslow several times for an idea of the number of "close species" in botanical genera (June 27, July 2, and July 7, 1855) before he managed to make clear that he was after an impression of how many almost indistinguishable species exist in large genera. Henslow apparently succeeded, for on July 21, 1855, he replied:

I thank you much for attempting to mark the list of dubious species: I was afraid it was a very difficult task, from, as you say, the want of a definition of what a species is.—I think however you were marking exactly what I wanted to know. My wish was derived as follows: I have ascertained, that APPARENTLY (I will not take up time by showing how) there is more variation, a wider geographical range, & probably more individuals, in the species of *large* genera than in the species of *small* genera. These general facts seem to me very curious, & I wanted to ascertain one point more; viz whether the closely allied and dubious forms which are generally considered as species, also belonged on average to large genera. [Barlow 1967: 182]

Note that here Darwin is also wrestling with the question of genera being real, a view he never entirely abandoned. In a letter to Asa Gray in 1857 (November 29), Darwin discusses his by-now established opinion that there is no clear distinction between varieties and species, and how there seems under Darwin's evolutionary views to be no easy foundation or set of physical criteria to decide when a variety has earned a specific epithet:

You speak of species not having any material base to rest on; but is this any greater hardship than deciding what deserves to be called a variety & be designated by a greek letter. When I was at systematic work, I know I longed to have no other difficulty (great enough) than deciding whether the form was distinct enough to deserve a name; & not to be haunted with undefined & unanswerable question whether it was a true species. What a jump it is from a well marked variety, produced by natural cause, to a species produced by the separate act of the Hand of God. But I am running on foolishly.—By the way I met the other day Phillips, the Palaeontologist,

& he asked me "how do you define a species?"—I answered "I cannot" Whereupon he said "at last I have found out the only true definition,—'any form which has ever had a specific name'"! [Burkhardt 1996: 183]

This anecdote found its way into later mythology in Poulton's essay on the species "problem" [Poulton 1903, 1908], although he dates it a week after the *Origin* (as discussed later). After the *Origin* was published, Darwin had sought Henslow's reaction[2] (November 11, 1859): "If you are *in even so slight a degree* staggered (which I hardly expect) on the immutability of species, then I am convinced with further reflection you will become more and more staggered, for this has been the process through which my mind has gone."

It seems Henslow was sufficiently staggered. He shortly afterward wondered at Owen's savage reaction to the views of the *Origin* (May 5, 1860):

When his own are to a certain extent of the same character. If I understand him, he thinks the "Becoming" of species (I suppose he means the *producing* of species) a somewhat rapid and not a slow process—but he seems to think them *progressive* organised [sic] out of previously organized beings {analogous (?) to minerals (simple and compound) out of ± 60 Elements}.

And when Sedgwick attacked Darwin in an address, Henslow defended him actively and forthrightly, also saying in his lectures to students, he reported, "[H]ow frequently Naturalists were at fault in regarding as *species*, forms which had (in some cases) been shown to be varieties, and how legitimately Darwin had deduced his *inferences* from positive experiment" [letter dated May 10, 1860, to Hooker, which was then passed on to Darwin].

In correspondence with Huxley before the *Origin*, on September 26, 1853, and October 3, 1853 (Padian 1999: 355), Darwin discussed the "Natural System" of classification: it was merely genealogical; we did not have access to a written record, and thus we had to work it out, but the cause of analogy and homology was genealogy (i.e., descent). Huxley replied that "Cuvier's definition of the object of Classification seems to me to embody all that is really wanted in Science—it is *to throw the facts of structure into the fewest possible general propositions*" [emphasis in original]. Darwin replied, "I knew, of course, of the Cuvierian view of Classification, but I think that most naturalists look for something further, & search for 'the natural system',—'for the plan on which the Creator has worked' &c &c.—It is this further element which I believe to be simply genealogical."

In summary, his pre-*Origin* correspondence shows him to be rather cautious about showing his hand, but he did seek information about what we would now call, following Mayr, "sibling species" and "cryptic species."

DARWIN'S PUBLISHED COMMENTS ON SPECIES BEFORE THE *ORIGIN*

In his *Journal of Researches* [Darwin 1839], Darwin makes few comments about species except to note, in the second edition of 1845, eight years after his thinking about transmutation began, that there are checks on the increases of populations.

> Every animal in a state of nature regularly breeds; yet in a species long established, any *great* increase in numbers is obviously impossible, and must be checked by some means. We are, nevertheless, seldom able with certainty to tell in any given species, at what period of life, or at what period of the year, or whether only at long intervals, the check falls; or, again, what is the precise nature of the check. Hence probably it is, that we feel so little surprise at one, of two species closely allied in habits, being rare and the other abundant in the same district; or, again, that one should be abundant in one district, and another, filling the same place in the economy of nature, should be abundant in a neighbouring district, differing very little in its conditions. If asked how this is, one immediately replies that it is determined by some slight difference, in climate, food, or the number of enemies: yet how rarely, if ever, we can point out the precise cause and manner of action of the check! We are, therefore, driven to the conclusion, that causes generally quite inappreciable by us, determine whether a given species shall be abundant or scanty in numbers. [*Journal of Researches* (Darwin 1845, Chapter VIII, p. 167)]

He also notes the difficulty of defining species in terms of morphology: "Of the latter [rabbit, a piebald hybrid of black and gray breeds] I now possess a specimen, and it is marked about the head differently from the French specific description. This circumstance shows how cautious naturalists should be in making species; for even Cuvier, on looking at the skull of one of these rabbits, thought it was probably distinct!" [Chapter IX, p. 184]. Darwin reports that the gauchos of South America were able to tell that these were one species because they shared the same territory and interbred, while Cuvier used only morphology. He says in a footnote, "The distinction of the rabbit as a species, is taken from peculiarities in the fur, from the shape of the head, and from the shortness of the ears. I may here observe that the difference between the Irish and English hare rests upon nearly similar characters, only more strongly marked" [p. 184n.]. From this we may conclude that Darwin was no naive morphologist, at any rate. He goes on to note that climate and ecotype are not the reason for species, nor is soil (contrary to Buffon and Trémaux, as discussed later), as nearly identical regions produce different species:

> I was much struck with the marked difference between the vegetation of these eastern valleys and those on the Chilian side: yet the climate, as well as the kind of soil, is nearly the same, and the difference of longitude very trifling. The same remark holds good with the quadrupeds, and in a lesser degree with the birds and insects. I may instance the mice, of which I obtained thirteen species on the shores of the Atlantic, and five on the Pacific, and not one of them is identical. We must except all those species, which habitually or occasionally frequent elevated mountains; and certain birds, which range as far south as the Strait of Magellan. This fact is in perfect accordance with the geological history of the Andes; for these mountains have existed as a great barrier since the present races of animals have appeared; and therefore, unless we suppose the same species to have been created in two different places, we ought not to expect any closer similarity between the organic beings on the opposite sides of the Andes than on the opposite shores of the ocean. In both cases, we must leave out of the question those kinds which have been able to cross the barrier, whether of solid rock or salt-water.[5]

> [5]This is merely an illustration of the admirable laws, first laid down by Mr. Lyell, on the geographical distribution of animals, as influenced by geological changes. The whole reasoning, of course, is founded on the assumption of the immutability of species; otherwise the difference in the species in the two regions might be considered as superinduced during a length of time. [Chapter XV, p. 313]

Here he is undercutting the widely held view that species are formed by climate or soil, a view that goes back to the medieval era (and motivated Buffon's view of deviation from the *première souche*). Instead, we have the beginnings of a biogeographic view of species as the result of geological isolation, which as we have seen in his Notebooks was a focus of his thinking at this time.

> Besides the several evident causes of destruction, there appears to be some more mysterious agency generally at work. Wherever the European has trod, death seems to pursue the aboriginal. We may look to the wide extent of the Americas, Polynesia, the Cape of Good Hope, and Australia, and we find the same result. Nor is it the white man alone that thus acts the destroyer; the Polynesian of Malay extraction has in parts of the East Indian

archipelago, thus driven before him the dark-coloured native. The varieties of man seem to act on each other in the same way as different species of animals—the stronger always extirpating the weaker. [Chapter XIX, p. 419]

Again, in hindsight we can see Darwin foreshadowing the idea that better-adapted species will exclude other species in competition with them.

In *Coral Reefs* [Darwin 1842], Darwin uses the term *species* conventionally, never noting any great problem with corals.[3] In the "Monograph on Cirripedia" [Darwin 1851], he describes his taxonomic practice in the preface:

In those cases in which a genus includes only a single species, I have followed the practice of some botanists, and given only the generic character, believing it to be impossible, before a second species is discovered, to know which characters will prove of specific, in contradistinction to generic, value.

In accordance with the Rules of the British Association, I have faithfully endeavoured to give to each species the first name attached to it, subsequently to the introduction of the binomial system, in 1758, in the tenth edition.[1] In accordance with the Rules, I have rejected all names before this date, and all MS. names. In one single instance, for reasons fully assigned in the proper place, I have broken through the great law of priority. I have given much fewer synonyms than is usual in conchological works; this partly arises from my conviction that giving references to works, in which there is not any original matter, or in which the Plates are not of a high order of excellence, is absolutely injurious to the progress of natural history, and partly, from the impossibility of feeling certain to which species the short descriptions given in most works are applicable;— thus, to take the commonest species, the *Lepas anatifera*, I have not found a single description (with the exception of the anatomical description by M. Martin St. Ange) by which this species can be certainly discriminated from the almost equally common *Lepas Hillii*. I have, however, been fortunate in having been permitted to examine a considerable number of authentically named specimens, (to which I have attached the sign (!) used by botanists,) so that several of my synonyms are certainly correct.

[1]In the Rules published by the British Association, the 12th edition, (1766,) is specified, but I am informed by Mr. Strickland that this is an error, and that the binomial method was followed in the 10th edition. of the 'Systema Naturæ.' [Part I, pp. ix–x]

Darwin is correct on both points. The Strickland Rules, as they came to be known, did indeed originally cite the twelfth edition, but the tenth was the edition in which bionomials were first used consistently and that became the benchmark edition. Darwin was a member of the commission that determined the Strickland Rules in 1842 [Amundson 2005: 47], which became the foundation for later strict taxonomic protocols. They were published and adopted by the British Association for the Advancement of Science, forming the basis for the later Blanchard Code of 1889, itself the basis for the International Rules of 1898, adopted in 1901 [Mayr, Linsley, and Usinger 1953: 205]. McOuat [1996] observes that this was part of a general professionalization of taxonomy, removing the "right" to name species from birdwatchers and gardeners to preclude confusion and synonymity. It is worth noting, with Amundson, that these were only nominally essentialistic—the name had to have a definition, but there was no requirement that the species taxon had an essence.

ON THE ORIGIN OF SPECIES, ON SPECIES

In the *Origin* [all quotations and page numbers from the sixth edition, Darwin 1872)],[4] Darwin makes many substantive and theoretical claims about species. In the chapter entitled "Variation under Domestication," he makes the following statements intended to convince those who rejected the mutability of species on logical grounds as well as practical ones. He notes that variation is an established fact within species, and that morphology is not a safe guide:

> Indefinite variability is a much more common result of changed conditions than definite variability, and has probably played a more important part in the formation of our domestic races. We see indefinite variability in the endless slight peculiarities which distinguish the individuals of the same species, and which cannot be accounted for by inheritance from either parent or from some more remote ancestor. [p. 6]

> Altogether at least a score of pigeons might be chosen, which, if shown to an ornithologist, and he were told that they were wild birds, would certainly be ranked by him as well-defined species. [p. 17]

> May not those naturalists who, knowing far less of the laws of inheritance than does the breeder, and knowing no more than he does of the intermediate links in the long lines of descent, yet admit that many of our domestic races are descended from the same parents—may they not learn a lesson of caution, when they deride the idea of species in a state of nature being lineal descendants of other species? [p. 21]

> But what concerns us is that the domestic varieties of the same species differ from each other in almost every character, which man has attended to and selected, more than do the distinct species of the same genera. [p. 31]

In the subsequent chapter, "Variation under Nature," he points out that there are several definitions of the notion of species, including many that involves special creation:

> Before applying the principles arrived at in the last chapter to organic beings in a state of nature, we must briefly discuss whether these latter are subject to any variation. To treat this subject properly, a long catalogue of dry facts ought to be given; but these I shall reserve for a future work. Nor shall I here discuss the various definitions which have been given of the term species. No one definition has satisfied all naturalists; yet every naturalist knows vaguely what he means when he speaks of a species. Generally the term includes the unknown element of a distinct act of creation. The term "variety" is almost equally difficult to define; but here community of descent is almost universally implied, though it can rarely be proved. We have also what are called monstrosities; but they graduate into varieties. By a monstrosity I presume is meant some considerable deviation of structure, generally injurious, or not useful to the species. Some authors use the term "variation" in a technical sense, as implying a modification directly due to the physical conditions of life; and "variations" in this sense are supposed not to be inherited; but who can say that the dwarfed condition of shells in the brackish waters of the Baltic, or dwarfed plants on Alpine summits, or the thicker fur of an animal from far northwards, would not in some cases be inherited for at least a few generations? And in this case I presume that the form would be called a variety. [p. 33]

There is, he says, a continuum of variation from occasional sports through to well-marked varieties, and some groups of organisms contain within them an enormous amount of variation:

> There is one point connected with individual differences, which is extremely perplexing: I refer to those genera which have been called "protean" or "polymorphic," in which the species present an inordinate amount of variation. With respect to many of these forms, hardly two naturalists agree whether to rank them as species or as varieties. We may instance Rubus, Rosa, and Hieracium amongst plants, several genera of insects and of Brachiopod shells. In most polymorphic genera some of the species have fixed and definite characters. Genera which are polymorphic in one country, seem to be, with a few exceptions, polymorphic in other countries, and likewise, judging from Brachiopod shells, at former periods of time. These facts are very perplexing, for they seem to show that this kind of variability is independent of the conditions of life. I am inclined to suspect that we see, at least in some of these polymorphic genera, variations which are of no service or disservice to the species, and which consequently have not been seized on and rendered definite by natural selection, as hereafter to be explained. [p. 35]

Purely morphological notions present difficulties to all naturalists; we find intermediate forms all the time, and inductively, Darwin is suggesting that this does not stop merely within species or between extant forms within genera:

> The forms which possess in some considerable degree the character of species, but which are so closely similar to other forms, or are so closely linked to them by intermediate gradations, that naturalists do not like to rank them as distinct species, are in several respects the most important for us. We have every reason to believe that many of these doubtful and closely allied forms have permanently retained their characters for a long time; for as long, as far as we know, as have good and true species. Practically, when a naturalist can unite by means of intermediate links any two forms, he treats the one as a variety of the other; ranking the most common, but sometimes the one first described, as the species, and the other as the variety. But cases of great difficulty, which I will not here enumerate, sometimes arise in deciding whether or not to rank one form as a variety of another, even when they are closely connected by intermediate links; nor will the commonly assumed hybrid nature of the intermediate forms always remove the difficulty. In very many cases, however, one form is ranked as a variety of another, not because the intermediate links have actually been found, but because analogy leads the observer to suppose either that they do now somewhere exist, or may formerly have existed; and here a wide door for the entry of doubt and conjecture is opened.
>
> Hence, in determining whether a form should be ranked as a species or a variety, the opinion of naturalists having sound judgment and wide experience seems the only guide to follow. We must, however, in many cases, decide by a majority of naturalists, for few well-marked and well-known varieties can be named which have not been ranked as species by at least some competent judges. [p. 36]

In short, then, there is often no consensus, and the facts have to be worked out by the most informed majority. Even then, there is often no fact of the matter when a variety is to be distinguished from a species. Darwin is undercutting the intuitions of his professional audience here.

> The geographical races or sub-species are local forms completely fixed and isolated; but as they do not differ from each other by strongly marked and important characters, "There is no possible test but individual opinion to determine which of them shall be considered as species and which as varieties" [quoting Wallace]. Lastly, representative species fill the same place in the natural economy of each island as do the local forms and sub-species; but as they are distinguished from each other by a greater amount of difference than that between the local forms and sub-species, they are almost universally ranked by naturalists as true

species. Nevertheless, no certain criterion can possibly be given by which variable forms, local forms, sub-species, and representative species can be recognised. [p. 38]

In fact, he says, sometimes the distinctness of species is due to the systematist classifying every variant as a distinct species (the splitters of modern taxonomy).

Some few naturalists maintain that animals never present varieties; but then these same naturalists rank the slightest difference as of specific value; and when the same identical form is met with in two distinct countries, or in two geological formations, they believe that two distinct species are hidden under the same dress. The term species thus comes to be a mere useless abstraction, implying and assuming a separate act of creation. It is certain that many forms, considered by highly competent judges to be varieties, resemble species so completely in character, that they have been thus ranked by other highly competent judges. But to discuss whether they ought to be called species or varieties, before any definition of these terms has been generally accepted, is vainly to beat the air. [p. 39]

I have been struck with the fact, that if any animal or plant in a state of nature be highly useful to man, or from any cause closely attracts his attention, varieties of it will almost universally be found recorded. These varieties, moreover, will often be ranked by some authors as species. Look at the common oak, how closely it has been studied; yet a German author makes more than a dozen species out of forms, which are almost universally considered by other botanists to be varieties; and in this country the highest botanical authorities and practical men can be quoted to show that the sessile and pedunculated oaks are either good and distinct species or mere varieties. [p. 40]

The ubiquitous variation of organisms, supposed by some to have been Darwin's major contribution to "population thinking" [Sober 1980; see Levit and Meister 2006], is something he derives from the work of Alphonse de Candolle:

I may here allude to a remarkable memoir lately published by A. [Alphonse] de Candolle, on the oaks of the whole world. No one ever had more ample materials for the discrimination of the species, or could have worked on them with more zeal and sagacity. . . . De Candolle then goes on to say that he gives the rank of species to the forms that differ by characters never varying on the same tree, and never found connected by intermediate states. After this discussion, the result of so much labour, he emphatically remarks: "They are mistaken, who repeat that the greater part of our species are clearly limited, and that the doubtful species are in a feeble minority. This seemed to be true, so long as a genus was imperfectly known, and its species were founded upon a few specimens, that is to say, were provisional. Just as we come to know them better, intermediate forms flow in, and doubts as to specific limits augment." [p. 40]

Intermediates are common, then, and claims of distinctness seem to rely on an a priori notion of descent from created parents, rather than be evidence in favor of the idea.

> When a young naturalist commences the study of a group of organisms quite unknown to him, he is at first much perplexed in determining what differences to consider as specific, and what as varietal; for he knows nothing of the amount and kind of variation to which the group is subject; and this shows, at least, how very generally there is some variation. But if he confine his attention to one class within one country, he will soon make up his mind how to rank most of the doubtful forms. His general tendency will be to make many species, for he will become impressed, just like the pigeon or poultry fancier before alluded to, with the amount of difference in the forms which he is continually studying; and he has little general knowledge of analogical variation in other groups and in other countries, by which to correct his first impressions. As he extends the range of his observations, he will meet with more cases of difficulty; for he will encounter a greater number of closely allied forms. But if his observations be widely extended, he will in the end generally be able to make up his own mind; but he will succeed in this at the expense of admitting much variation, and the truth of this admission will often be disputed by other naturalists. When he comes to study allied forms brought from countries not now continuous, in which case he cannot hope to find intermediate links, he will be compelled to trust almost entirely to analogy, and his difficulties will rise to a climax.
>
> Certainly no clear line of demarcation has as yet been drawn between species and sub-species—that is, the forms which in the opinion of some naturalists come very near to, but do not quite arrive at, the rank of species: or, again, between sub-species and well-marked varieties, or between lesser varieties and individual differences. These differences blend into each other by an insensible series; and a series impresses the mind with the idea of an actual passage. [p. 41]

This passage is most critical—Darwin has moved from the formal variation of groups to the idea that these forms are the result of a temporal sequence. Moreover, the only difference between slight variants, marked variations, and species is a matter of time passed. The difference of *rank* is arbitrary:

> From these remarks it will be seen that I look at the term species as one arbitrarily given, for the sake of convenience, to a set of individuals closely resembling each other, and that it does not essentially differ from the term

variety, which is given to less distinct and more fluctuating forms. The term variety, again, in comparison with mere individual differences, is also applied arbitrarily, for convenience' sake. [p. 42]

We should be cautious here. It does not seem to me that Darwin is saying that the *groupings* are arbitrary, and the Notebook comment that species are real to their contemporaries backs up the claim that he thinks the groups are natural. What he thinks is arbitrary is where the distinction between the ranks of *species* and *variety* is to be drawn. And genera are the result of the age since common parenthood; some are larger than others and have more variation because the conditions that cause species to form have remained favorable for a long time:

From looking at species as only strongly marked and well-defined varieties, I was led to anticipate that the species of the larger genera in each country would oftener present varieties, than the species of the smaller genera; for wherever many closely related species (i.e., species of the same genus) have been formed, many varieties or incipient species ought, as a general rule, to be now forming. Where many large trees grow, we expect to find saplings. Where many species of a genus have been formed through variation, circumstances have been favourable for variation; and hence we might expect that the circumstances would generally be still favourable to variation. On the other hand, if we look at each species as a special act of creation, there is no apparent reason why more varieties should occur in a group having many species, than in one having few. [p. 44]

Moreover, the species of the larger genera are related to each other in the same manner as the varieties of any one species are related to each other. No naturalist pretends that all the species of a genus are equally distinct from each other; they may generally be divided into sub-genera, or sections, or lesser groups. As Fries has well remarked, little groups of species are generally clustered like satellites around other species. And what are varieties but groups of forms, unequally related to each other, and clustered round certain forms—that is, round their parent-species? Undoubtedly there is one most important point of difference between varieties and species; namely, that the amount of difference between varieties, when compared with each other or with their parent-species, is much less than that between the species of the same genus. [p. 46]

So Darwin presents species as real, but not as a formal and fixed rank. Genera are like species and varieties in that they are the result of groupings of variation. They, too, do not seem to be a fixed rank, commensurate across all genera. He summarizes the argument in this chapter as follows:

Finally, varieties cannot be distinguished from species,—except, first, by the discovery of intermediate linking forms; and, secondly, by a certain indefinite amount of difference between them; for two forms, if differing very little, are generally ranked as varieties, notwithstanding that they cannot be closely connected; but the amount of difference considered necessary to give to any two forms the rank of species cannot be defined. In genera having more than the average number of species in any country, the species of these genera have more than the average number of varieties. In large genera the species are apt to be closely, but unequally, allied together, forming little clusters round other species. Species very closely allied to other species apparently have restricted ranges. In all these respects the species of large genera present a strong analogy with varieties. And we can clearly understand these analogies, if species once existed as varieties, and thus originated; whereas, these analogies are utterly inexplicable if species are independent creations. [p. 47]

Darwin had no doubt that species were formed through selection on varietal forms, and this provides the missing mechanism for how conditions of life can give rise to varieties:

Again, it may be asked, how is it that varieties, which I have called incipient species, become ultimately converted into good and distinct species, which in most cases obviously differ from each other far more than do the varieties of the same species? How do those groups of species, which constitute what are called distinct genera, and which differ from each other more than do the species of the same genus, arise? All these results, as we shall more fully see in the next chapter, follow from the struggle for life. Owing to this struggle, variations, however slight and from whatever cause proceeding, if they be in any degree profitable to the individuals of a species, in their infinitely complex relations to other organic beings and to their physical conditions of life, will tend to the preservation of such individuals, and will generally be inherited by the offspring. The offspring, also, will thus have a better chance of surviving, for, of the many individuals of any species which are periodically born, but a small number can survive. I have called this principle, by which each slight variation, if useful, is preserved, by the term Natural Selection, in order to mark its relation to man's power of selection. [Chapter III, p. 48f.]

Selection is for Darwin not restricted to the intraspecific but can be interspecific or even merely a matter of simple survival:

Hence, as more individuals are produced than can possibly survive, there must in every case be a struggle for existence, either one individual with another of the same species, or with the individuals of distinct species, or with the physical conditions of life. [p. 50]

But the struggle will almost invariably be most severe between the individuals of the same species, for they frequent the same districts, require the same food, and are exposed to the same dangers. In the case of varieties of the same species, the struggle will generally be almost equally severe, and we sometimes see the contest soon decided. [p. 58f.]

And in Chapter IV, he notes, "In order that any great amount of modification should be effected in a species, a variety, when once formed, must again, perhaps after a long interval of time, vary or present individual differences of the same favourable nature as before; and these must be again preserved, and so onwards step by step" [p. 66].

Darwin had no trouble finding links between asexuals and self-fertilizing species, and it is clear that he did not exclude asexuals from *being* species, as we see:

> It must have struck most naturalists as a strange anomaly that, both with animals and plants, some species of the same family and even of the same genus, though agreeing closely with each other in their whole organisation, are hermaphrodites, and some unisexual. But if, in fact, all hermaphrodites do occasionally intercross, the difference between them and unisexual species is, as far as function is concerned, very small. [p. 79]

Darwin defined the processes that keep species distinct in even occasionally sexual organisms as the result of intercrossing. The benefits of sexual reproduction include "vigour and fertility," both germane to selection (although he doesn't explain exactly why, which resulted in extensive and ongoing debates in the following century on the evolutionary benefits of sex). But equally interesting here is that, contrary to many twentieth-century Darwinians (for example, Fisher and Simpson, discussed later), Darwin himself has no problem explaining asexual species, and even explains them, as Manfred Eigen does today [Eigen 1993b], as the result of natural selection entirely.

> Intercrossing plays a very important part in nature by keeping the individuals of the same species, or of the same variety, true and uniform in character. It will obviously thus act far more efficiently with those animals which unite for each birth; but, as already stated, we have reason to believe that occasional intercrosses take place with all animals and plants. Even if these take place only at long intervals of time, the young thus produced will gain so much in vigour and fertility over the offspring from long-continued self-fertilisation, that they will have a better chance of surviving and propagating their kind; and thus in the long run the influence of crosses, even at rare intervals, will be great. With respect to organic beings extremely low in the scale, which do not propagate sexually, nor conjugate, and which cannot possibly intercross, uniformity of character can be retained by them under the same conditions of life, only through the principle of inheritance, and

through natural selection which will destroy any individuals departing from the proper type. If the conditions of life change and the form undergoes modification, uniformity of character can be given to the modified offspring, solely by natural selection preserving similar favourable variations. [p. 81]

Darwin at the time of the sixth edition thought that the formation of species did not rely on isolation, and here he takes Moritz Wagner's view to task [Wagner 1889]. Isolation does, he thought, make species formation easier, but he cannot agree it is required, as Wagner thought. And if the isolated population is *too* small, then isolation can in fact *prevent* speciation from occurring due to a lack of variation. The founder effect or drift through biased sampling has not occurred to him, as it later did to Weismann.[5]

Isolation, also, is an important element in the modification of species through natural selection. In a confined or isolated area, if not very large, the organic and inorganic conditions of life will generally be almost uniform; so that natural selection will tend to modify all the varying individuals of the same species in the same manner. Intercrossing with the inhabitants of the surrounding districts will, also, be thus prevented. Moritz Wagner has lately published an interesting essay on this subject, and has shown that the service rendered by isolation in preventing crosses between newly-formed varieties is probably greater even than I supposed. But from reasons already assigned I can by no means agree with this naturalist, that migration and isolation are necessary elements for the formation of new species. The importance of isolation is likewise great in preventing, after any physical change in the conditions such as of climate, elevation of the land, &c., the immigration of better adapted organisms; and thus new places in the natural economy of the district will be left open to be filled up by the modification of the old inhabitants. Lastly, isolation will give time for a new variety to be improved at a slow rate; and this may sometimes be of much importance. If, however, an isolated area be very small, either from being surrounded by barriers, or from having very peculiar physical conditions, the total number of the inhabitants will be small; and this will retard the production of new species through natural selection, by decreasing the chances of favourable variations arising. [p. 81f.]

Although isolation is of great importance in the production of new species, on the whole I am inclined to believe that largeness of area is still more important, especially for the production of species which shall prove capable of enduring for a long period, and of spreading widely. Throughout a great and open area, not only will there be a better chance of favourable variations, arising from the large number of individuals of the same species there supported, but the conditions of life are much more complex from the large number of already existing species; and if some of these many species become modified and improved, others will have to be improved in a corresponding degree, or they will be exterminated. Each new form, also, as soon as it has been much improved, will be able to spread over the open and continuous area, and will thus come into competition with many other forms. Moreover, great areas, though now continuous, will often, owing to former oscillations of level, have existed in a broken condition; so that the good effects of isolation will generally, to a certain extent, have concurred. Finally, I conclude that, although small isolated areas have been in some respects highly favourable for the production of new species, yet that the course of modification will generally have been more rapid on large areas; and what is more important, that the new forms produced on large areas, which already have been victorious over many competitors, will be those that will spread most widely, and will give rise to the greatest number of new varieties and species. They will thus play a more important part in the changing history of the organic world. [p. 82f.]

Ironically, as Kottler [1978: 285–288] noted, Darwin in the Notebooks believed that isolation *was* a *sine qua non* for speciation, in part following the views of Leopold von Buch. But by this later stage, Darwin appears to have made natural selection the primary cause of species, requiring that variation needs to occur *in situ* as it were, and so needing larger populations to give it opportunity to do so.

He also notes that the structures that are useful in classifying species are often not of any adaptive value: "Hence modifications of structure, viewed by systematists as of high value, may be wholly due to the laws of variation and correlation, without being, as far as we can judge, of the slightest service to the species" [p. 116].

Considering the traditional logical notions of generic and specific characters *(per genus et differentiam)* from which the concept of *species* was drawn by natural historians in the seventeenth century, it is interesting to note that in contrast, Darwin expects that the specific characters will vary more than the generic and, moreover, that if a character is variable in the genus between closely allied species, it is also more variable in the individual species as well:

[O]n the view that species are only strongly marked and fixed varieties, we might expect often to find them still continuing to vary in those parts of their structure which have varied within a moderately recent period, and which have thus come to differ. Or to state the case in another manner: the points in which all the species of a genus resemble each other, and in which they differ from allied genera, are called generic characters; and these characters may be attributed to inheritance from a common progenitor, for it can rarely have happened that natural selection will have modified several distinct species, fitted to more or less widely different habits, in exactly the same manner:—and as these so-called generic characters have been inherited from before the period when the several species first branched off from their common progenitor, and subsequently have not varied or come to differ in any degree, or only in a slight degree, it is not probable that they should vary at the present day. On the other hand, the points in which species differ from other species of the same genus are called specific characters; and as these specific characters have varied and come to differ since the period when the species branched off from a common progenitor, it is probable that they should still often be in some degree variable,—at least more variable than those parts of the organisation which have for a very long period remained constant. [p. 122f.]

Asking, in effect, why the Great Chain of Being principle of *lex completio* [Lovejoy 1936] does not result in no species being seen at all but instead one variable mass, Darwin answers, "I believe that species come to be tolerably well-defined objects, and do not at any one period present an inextricable chaos of varying and intermediate links" [p. 137], due to

(i)   the fact that variation and selection take time, and may not have yet occurred;

(ii) when smaller isolated populations spread out, selection exterminates the older intermediate forms (in modern terms, when in sympatry, allopatric variations exclude the less fit older forms);

(iii) intermediates are subject to accidental extinction because they are less widely spread; and

(iv) numberless intermediate varieties, linking closely together all the species of the same group, must assuredly have existed; but the very process of natural selection constantly tends, as has been so often remarked, to exterminate the parent-forms and the intermediate links. [p. 138]

When reading the *Origin*, one is struck by Darwin's repeated locutions "to the benefit of the species" or "of advantage to the species." Darwin seems to be making the (now) classical blunder of group selectionism. However, as one reads these examples, it becomes clear that Darwin is using it as a circumlocution for "of advantage to the members of the species that carry this trait." It is clear that he thinks that selection occurs, in the main, through competition between varieties (and of course species are just well-marked varieties, as he has said), increasing or decreasing in relative numbers as they carry beneficial or unbeneficial traits. In this usage, *species* is just a way of marking the variety that is subjected to selection or has been so subjected and gone to fixation; it is roughly equivalent, therefore, in Darwin's mind, to the sum total of what G. C. Williams [1966] later called "evolutionary genes"—those hereditable variations that are selectable.

But Darwin did not think that selection *caused* sterility and hence species. Rather, he held that selection incidentally resulted in forms that were unable to breed together:

> The view commonly entertained by naturalists is that species, when inter-crossed, have been specially endowed with sterility, in order to prevent their confusion. This view certainly seems at first highly probable, for species living together could hardly have been kept distinct had they been capable of freely crossing. The subject is in many ways important for us, more especially as the sterility of species when first crossed, and that of their hybrid off-spring, cannot have been acquired, as I shall show, by the preservation of successive profitable degrees of sterility. It is an incidental result of differences in the reproductive systems of the parent-species. [p. 235]

> The fertility of varieties, that is of the forms known or believed to be descended from common parents, when crossed, and likewise the fertility of their mongrel offspring, is, with reference to my theory, of equal importance with the sterility of species; for it seems to make a broad and clear distinction between varieties and species. [p. 236]

However, on examination, he does not think this holds generally true:

> It is certain, on the one hand, that the sterility of various species when crossed is so different in degree and graduates away so insensibly, and, on the other hand, that the fertility of pure species is so easily affected by various circumstances, that for all practical purposes it is most difficult to say where perfect fertility ends and sterility begins. I think no better evidence of this can be required than that the two most experienced observers who have ever lived, namely Kölreuter and Gärtner, arrived at diametrically opposite conclusions in regard to some of the very same forms. It is also most instructive to compare—but I have not space here to enter into details—the evidence advanced by our best botanists on the question whether certain doubtful forms should be ranked as species or varieties, with the evidence from fertility adduced by different hybridisers, or by the same observer from experiments made during different years. It can thus be shown that neither sterility nor fertility affords any certain distinction between species and varieties. The evidence from this source graduates away, and is doubtful in the same degree as is the evidence derived from other constitutional and structural differences. [p. 236f.]

Also:

> By the term systematic affinity is meant, the general resemblance between species in structure and constitution. Now the fertility of first crosses, and of the hybrids produced from them, is largely governed by their systematic

affinity. This is clearly shown by hybrids never having been raised between species ranked by systematists in distinct families; and on the other hand, by very closely allied species generally uniting with facility. But the correspondence between systematic affinity and the facility of crossing is by no means strict. A multitude of cases could be given of very closely allied species which will not unite, or only with extreme difficulty; and on the other hand of very distinct species which unite with the utmost facility. [p. 242f.]

And, reiterating the comment in the Notebook: "No one has been able to point out what kind or what amount of difference, in any recognisable character, is sufficient to prevent two species crossing" [p. 243].

Darwin notes that species are almost always intersterile but that this is often due to the fact that as soon as this intersterility is noticed, taxonomists will rank these varieties as species:

It may be urged, as an overwhelming argument, that there must be some essential distinction between species and varieties, inasmuch as the latter, however much they may differ from each other in external appearance, cross with perfect facility, and yield perfectly fertile offspring. With some exceptions, presently to be given, I fully admit that this is the rule. But the subject is surrounded by difficulties, for, looking to varieties produced under nature, if two forms hitherto reputed to be varieties be found in any degree sterile together, they are at once ranked by most naturalists as species. [p. 256]

Anyway, some obvious varieties within species are sterile together, even though they are able to breed with mutual races:

From these facts it can no longer be maintained that varieties when crossed are invariably quite fertile. From the great difficulty of ascertaining the infertility of varieties in a state of nature, for a supposed variety, if proved to be infertile in any degree, would almost universally be ranked as a species;— from man attending only to external characters in his domestic varieties, and from such varieties not having been exposed for very long periods to uniform conditions of life;—from these several considerations we may conclude that fertility does not constitute a fundamental distinction between varieties and species when crossed. The general sterility of crossed species may safely be looked at, not as a special acquirement or endowment, but as incidental on changes of an unknown nature in their sexual elements. [p. 259]

So, in the end, Darwin refuses to make sterility a test of species or even to expect that sterility will correlate with systematic affinity, summarizing the arguments in that chapter as follows:

First crosses between forms, sufficiently distinct to be ranked as species, and their hybrids, are very generally, but not universally, sterile. The sterility is of all degrees, and is often so slight that the most careful experimentalists have arrived at diametrically opposite conclusions in ranking forms by this test. The sterility is innately variable in individuals of the same species, and is eminently susceptible to the action of favourable and unfavourable conditions. The degree of sterility does not strictly follow systematic affinity, but is governed by several curious and complex laws. It is generally different, and sometimes widely different in reciprocal crosses between the same two species. It is not always equal in degree in a first cross and in the hybrids produced from this cross.

In the same manner as in grafting trees, the capacity in one species or variety to take on another, is incidental on differences, generally of an unknown nature, in their vegetative systems, so in crossing, the greater or less facility of one species to unite with another is incidental on unknown differences in their reproductive systems. There is no more reason to think that species have been specially endowed with various degrees of sterility to prevent their crossing and blending in nature, than to think that trees have been specially endowed with various and somewhat analogous degrees of difficulty in being grafted together in order to prevent their inarching in our forests. [p. 262]

In the chapter on classification, Darwin attended to the problems that this view brings with it for working naturalists but also the problems it solves, especially in understanding the reason for the systematic affinities:

Naturalists, as we have seen, try to arrange the species, genera, and families in each class, on what is called the Natural System. But what is meant by this system? Some authors look at it merely as a scheme for arranging together those living objects which are most alike, and for separating those which are most unlike; or as an artificial method of enunciating, as briefly as possible, general propositions,—that is, by one sentence to give the characters common, for instance, to all mammals, by another those common to all carnivora, by another those common to the dog-genus, and then, by adding a single sentence, a full description is given of each kind of dog. The ingenuity and utility of this system are indisputable. But many naturalists think that something more is meant by the Natural System; they believe that it reveals the plan of the Creator; that unless it be specified whether order in time or space, or both, or what else is meant by the plan of the Creator, it seems to me that nothing is thus added to our knowledge. Expressions such as that famous one by Linnæus, which we often meet with in a more or less concealed form, namely, that the characters do not make the genus, but that the genus gives the characters, seem to imply that some deeper bond is included in our classifications than mere resemblance. I believe that this is the case, and that community of descent—the one known cause of close similarity in organic beings—is the bond, which though observed by various degrees of modification, is partially revealed to us by our classifications. [p. 364f.]

The importance, for classification, of trifling characters, mainly depends on their being correlated with many other characters of more or less importance. The value indeed of an aggregate of characters is very evident in natural history. Hence, as has often been remarked, a species may depart from its allies in several characters, both of high physiological importance, and of almost universal prevalence, and yet leave us in no doubt where it should be ranked. Hence, also, it has been found that a classification founded on any single character, however important that may be, has always failed; for no part of the organisation is invariably constant. The importance of an aggregate of characters, even when none are important, alone explains the aphorism enunciated by Linnæus, namely, that the characters do not give the genus, but the genus gives the characters; for this seems founded on the appreciation of many trifling points of resemblance, to slight to be defined. [p. 367]

All the foregoing rules and aids and difficulties in classification may be explained, if I do not greatly deceive myself, on the view that the Natural System is founded on descent with modification;—that the characters which naturalists consider as showing true affinity between any two or more species, are those which have been inherited from a common parent, all true classification being genealogical;—that community of descent is the hidden bond which naturalists have been unconsciously seeking, and not some unknown plan of creation, or the enunciation of general propositions, and the mere putting together and separating objects more or less alike.

But I must explain my meaning more fully. I believe that the *arrangement* of the groups within each class, in due subordination and relation to each other, must be strictly genealogical in order to be natural; but that the *amount* of difference in the several branches or groups, though allied in the same degree in blood to their common progenitor, may differ greatly, being due to the different degrees of modification which they have undergone; and this is expressed by the forms being ranked under different genera, families, sections, or orders. [p. 369]

Given that there is some continuing dispute over whether or not Darwin was a cladist [Mayr 1982: 209–213; 1994], it is worth stating here my view that Padian [1999] is right—for Darwin in this last passage, classification is necessarily natural only if it matches genealogy, but it may also be represented in terms of grade of organization. If it is, though, such grades must not trim away the genealogical relationships. Darwin was not a cladist,[6] but he was pretty close to it. However, he recognizes the practical necessity of representing overall differences in a classification scheme. It's just not the same as saying these differences are part of a *natural* classification; this point shall become significant in the argument presented in the final chapters. But even if Darwin were an "eclectic" in his approach to classification, as Mayr suggests, this would be due to the fact that he had not yet completely worked out the implications of his view of species related by common descent.

Darwin then summarizes the difference between the process by which things have evolved and the patterns that diagnose them—his argument is remarkably similar to the view of the so-called father of cladistics, Willi Hennig, on "reciprocal illumination," in which he argues that knowledge of groups illuminate the uncovering of the knowledge of other groups, which then help refine the initial groups [Hennig 1966: 21f., 148, 206, 222]:

> With species in a state of nature, every naturalist has in fact brought descent into his classification; for he includes in his lowest grade, that of species, the two sexes; and how enormously these sometimes differ in the most important characters, is known to every naturalist: scarcely a single fact can be predicated in common of the adult males and hermaphrodites of certain cirripedes, and yet no one dreams of separating them. . . . The naturalist includes as one species the various larval stages of the same individual, however much they may differ from each other and from the adult, as well as the so-called alternate generations of Steenstrup, which can only in a technical sense be considered as the same individual. He includes monsters and varieties, not from their partial resemblance to the parent-form, but because they are descended from it.
>
> As descent has universally been used in classing together the individuals of the same species, though the males and females and larvæ are sometimes extremely different; and as it has been used in classing varieties which have undergone a certain, and sometimes a considerable, amount of modification, may not this same element of descent have been unconsciously used in grouping species under genera, and genera under higher groups, all under the so-called natural system? I believe it has been unconsciously used; and thus only can I understand the several rules and guides which have been followed by our best systematists. As we have no written pedigrees, we are forced to trace community of descent by resemblances of any kind. Therefore we choose those characters which are the least likely to have been modified, in relation to the conditions of life to which each species has been recently exposed. Rudimentary structures on this view are as good as, or even sometimes better than, other parts of the organisation. We care not how trifling a character may be—let it be the mere inflection of the angle of the jaw, the manner in which an insect's wing is folded, whether the skin be covered by hair or feathers—if it prevail throughout many and different species, especially those having very different habits of life, it assumes high value; for we can account for its presence in so many forms with such different habits, only by inheritance from a common parent. We may err in this respect in regard to single points of structure, but when several characters, let them be ever so trifling, concur throughout a large group of beings having different habits, we may feel almost sure, on the theory of descent, that these characters have been inherited from a common ancestor;

and we know that such aggregated characters have especial value in classification.

We can understand why a species or a group of species may depart from its allies, in several of its most important characteristics, and yet be safely classed with them. This may be safely done, and is often done, as along as a sufficient number of characters, let them be ever so unimportant, betray the hidden bond of community of descent. Let two forms have not a single character in common, yet, if these extreme forms are connected together by a chain of intermediate groups, we may at once infer their community of descent, and we put them all into the same class. As we find organs of high physiological importance—those which serve to preserve life under the most diverse conditions of existence—are generally the most constant, we attach especial value to them; but if these same organs, in another group or section of a group, are found to differ much, we at once value them less in our classification. We shall presently see why embryological characters are of such high classificatory importance. Geographical distribution may sometimes be brought usefully into play in classing large genera, because all the species of the same genus, inhabiting any distinct and isolated region, are in all probability descended from the same parents. [p. 372f.]

In the final chapter, the "Recapitulation and Conclusion," Darwin makes the now-famous comment about the reality of species that led so many to conclude he was a mere nominalist:

When the views advanced by me in this volume, and by Mr. Wallace, or when analogous views on the origin of species are generally admitted, we can dimly foresee that there will be a considerable revolution in natural history. Systematists will be able to pursue their labours as at present; but they will not be incessantly haunted by the shadowy doubt whether this or that form be a true species. This, I feel sure and I speak after experience, will be no slight relief. The endless disputes whether or not some fifty species of British brambles are good species will cease. Systematists will have only to decide (not that this will be easy) whether any form be sufficiently constant and distinct from other forms, to be capable of definition; and if definable, whether the differences be sufficiently important to deserve a specific name. This latter point will become a far more essential consideration than it is at present; for differences, however slight, between any two forms, if not blended by intermediate gradations, are looked at by most naturalists as sufficient to raise both forms to the rank of species.

Hereafter we shall be compelled to acknowledge that the only distinction between species and well-marked varieties is, that the latter are known, or believed, to be connected at the present day by intermediate gradations whereas species were formerly thus connected. Hence, without rejecting the consideration of the present existence of intermediate gradations between any two forms, we shall be led to weigh more carefully and to value higher the actual amount of difference between them. It is quite possible that forms now generally acknowledged to be merely varieties may hereafter be thought worthy of specific names; and in this case scientific and common language will come into accordance. In short, we shall have to treat species in the same manner as those naturalists treat genera, who admit that genera are merely artificial combinations made for convenience. This may not be a cheering prospect; but we shall at least be freed from the vain search for the undiscovered and undiscoverable essence of the term species. [p. 425f.]

It will be worthwhile to look at this matter closely, given the excerpts cited here as cross bearings on the subtlety of Darwin's views on species, and ask, Has Darwin said that species are "merely artificial combinations made for convenience"? The answer is no. He has said, rather, that naturalists shall be forced to treat them that way. We have seen repeatedly that Darwin did not insist *species* were unreal, merely that the *rank* was arbitrarily assigned and that we could not see if they were real on the basis of characters. The reason why the term *species* has no discoverable essence (but does that imply it has a Real Essence in the Lockean sense?) is that each case is different in the biological particulars. But they are separated, he says, in that the "intergradations" between them are extinct. *That* is real enough. Darwin's definition of species is simply that they do not interbreed or, in the case of "unisexual" organisms, that natural selection keeps them isolated in the "proper type" suited to the conditions of life in which they live. In this I am concurring with Kottler's and Ghiselin's conclusions; where Darwin seems to be a nominalist, he is in fact describing the problems of current taxonomic criteria, founded on creationist views of species, and so is describing what species are *not* [Kottler 1978: 293f.; Ghiselin 1984].

The glossary to the later editions of the *Origin* does not give us a definition for species, so we lack from Darwin the sort of epigrammatic slogan for species that so many other writers have given us and even if it had, it would be the definition of the compiler W. S. Dallas, not of Darwin. David Williams notes (personal communication) that Dallas's definition of terms like *homology* differed from Darwin's use of the term.

AFTER THE *ORIGIN*

After the *Origin*, Darwin is able to publish some generalizations about species. He does so in chapter 2 of the *Descent of Man* [2nd ed., Darwin 1871]:

If we consider all the races of man as forming a single species, his range is enormous; but some separate races, as the Americans and Polynesians, have very wide ranges. It is a well-known law that widely-ranging species are much more variable than species with restricted ranges; and the variability of man may with more truth be compared with that of widely-ranging species, than with that of domesticated animals. [p. 29]

The "well-known law" is in part of Darwin's own construction, as we have seen.

In the *Variation of Plants and Animals under Domestication* [Darwin 1998], published in 1868 and revised in 1875, Darwin discusses how species arise in nature and proposes intersterility as a criterion of species status.

> [H]ow, it may be asked, have species arisen in a state of nature? The differences between natural varieties are slight; whereas the differences are considerable between the species of the same genus, and great between the species of distinct genera. How do these lesser differences become augmented into the greater difference? How do varieties, or as I have called them, incipient species, become converted into true and well-defined species? [Volume I, p. 5]

Using domestic varieties as a guide to variation in nature, as he had in the *Origin*, he notes "that the sterility of distinct species when crossed, and of their hybrid progeny, depends exclusively on the nature of their sexual elements, and not on any differences in their structure or general constitution. . . . That excellent observer, Gärtner, likewise concluded that species when crossed are sterile owing to differences confined to their reproductive systems" [Volume II, p. 168f.]. And in answer to the question of how it is that extremely different domesticated varieties are "perfectly fertile" while closely allied species are not, he responds, "Passing over the fact that the amount of external difference between two species is no sure guide to the degree of their mutual sterility, so that similar differences in the case of varieties [of domestic animals and plants—JSW] would be no sure guide, we know that with species the cause lies exclusively in differences in their sexual constitution" [Volume II, p. 172].

In a short letter to *Nature* [Darwin 1873], while discussing whether variations in brain structure could enable the evolution of instincts, Darwin notes in his own defense:

> The writer of the article in referring to my words "the preservation of useful variations of pre-existing instincts" adds "the question is, whence these variations?" Nothing is more to be desired in natural history than that some one would be able to answer such a query. But as far as our present subject is concerned, the writer probably will admit that a multitude of variations have arisen, for instance in colour and in the character of the hair, feathers, horns, &c., which are quite independent of habit and of use in previous generations. It seems far from wonderful, considering the

complex conditions to which the whole organisation is exposed during the successive stages of its development from the germ, that every part should be liable to occasional modifications: the wonder indeed is that any two individuals of the same species are at all closely alike.

So, in the end, Darwin proposed a "snowflake" theory of species: all members are alike in some ways, but they are also unique individuals. Variation occurs naturally, and it is weeded out according to how well suited it is to the conditions of life. Species are held distinct incidentally to their being adapted to those conditions; they are real at the time, although no rank seems to be absolute, and there is no particular amount or kind of difference between them that marks out, or correlates with being, distinct species. For sexual organisms all that can be said is that they do not, in nature, interbreed (no matter whether they can be made to in captivity). He is not a nominalist, as he notes in his response to Agassiz's criticism: "[I]f species do not exist at all, as the supporters of the transmutation theory maintain, how can they vary? And if individuals alone exist, how can differences which may be observed among them prove the variability of species?" [in Lurie 1960: 297]. Darwin's reply, to Asa Gray, was "I am surprised that Agassiz did not succeed in writing something better. How absurd that logical quibble 'if species do not exist how can they vary'? As if anyone doubted their temporary existence?" [quoted in Gayon 1996: 229; cf. also Ghiselin 1969, chapter 4].

We need to note with Darwin, as with others, that there is a difference between denying that the *rank* of species has a definition and denying that the *term species* has one. Darwin denies the former, but not the latter. He is not a nominalist but a pluralist with regard to what makes species distinct. Nevertheless, all these causes resolve down to an aversion to interbreeding in sexual organisms and differences in their sexual structures and constitutions, and selection maintaining the appropriate forms and organs for living in the conditions in which they find themselves, for asexuals. Of course, he also allowed that conditions of life may directly affect both the sexual organs and the structures of the organisms [e.g., Darwin 1998, Volume II, p. 413]. Nevertheless, I think we can dispose of Futuyma's and others' mischaracterization. Darwin knew very well how to define species, given the fact of evolution.

In summary, Darwin was a species realist but denied the absolute rank of Linnaean classification, although he used it in practice and was a contributor to the Strickland Rules. He allowed for asexual species, and his views focused primarily on the process of speciation, which he initially entertained might be due to geographic isolation (and possibly the effects of the local climate, as in the traditional view) but later held was due to the fixing of subspecific varieties due to natural selection. Speciation is a side effect of selection for varieties, which causes changes in the sexual reproduction system of sexual species.[7]

MORITZ WAGNER, PIERRE TRÉMAUX, AND GEOGRAPHIC SPECIATION

Mortiz Wagner (1813–1887) was a celebrated explorer and geographer, and his writings were influential. He had proposed, in opposition to Darwin's notion that species are formed from racial types through selection, that species must be isolated geographically [Wagner 1889]. Mayr [1982: 562–566] discusses the reaction of Darwin and his contemporaries to Wagner's isolation model. Dismissed by Weismann and Wallace, as we shall see, the geographic isolation thesis was nevertheless adopted by the Reverend Gulick, whose work on Hawaiian lands nails (genus *Achatinella*) led him also to claim that much evolutionary variation was due to chance [Amundson 1996; Mayr 1982: 555]. Thereafter, there seemed to be two camps: those who thought that isolation was the *sine qua non* of speciation and that chance was the cause of variation between populations, and those who thought that speciation occurred equally if not entirely through the action of natural and perhaps sexual selection. After the later work of Sewall Wright had been accepted and promoted in the work of Dobzhansky during the mid–twentieth century, and following Poulton's and Mayr's coinages, this view came to be known as the *allopatric* theory of speciation, and Darwin's published view as the *sympatric* theory [Depew and Weber 1995: 275–278].

A possible source for the notion of geographic isolation as a mechanism for speciation is the work of the amateur anthropologist and architecture scholar, Pierre Trémaux, whose 1865 work *Origin et transformations de l'homme et des autres êtres* (The Origin and Evolution of Man and Other Beings) proposed that the effects of local climate and conditions *(sol)* would fully determine racial and species characters by adaptation, which would then be maintained by interbreeding or, as he calls it, "crossing" *(croisement)*. Such changes would happen rapidly, evolutionarily speaking, and then reach an *equilibrium*. Charitably interpreted, for the language of *Origin et transformations* is florid and nontechnical, Trémaux proposed both an allopatric theory and a punctuated equilibrium theory. However, his amateur standing in the French scientific community meant that his self-published works were not taken seriously, although Darwin had two copies in his library and had read at least one of them [Wilkins and Nelson 2008], and as a result may have revised his expression on pages 409f. of the 1866 fourth edition of the *Origin of Species* regarding the rates of change to include more or less rapid changes. Trémaux has been unjustly tarred with being a crank because Marx wrote to Engels to recommend him over Darwin, and Engels replied that Trémaux was simply silly and ignorant of the facts. Nearly every commentator since has relied on Engels's interpretation,[8] which is to my mind clearly a strawman interpretation based on Engels's own ignorance of the use of the French term *sol* by Buffon and others to mean "habitat." In his 1865 work, there is an extended discussion of the definitions of "species," the first such discussion I know of, mostly focusing on French definitions from Cuvier. It is possible, although there is no direct evidence as yet, that Wagner also had read Trémaux's ideas when he published eight years afterward. Trémaux later published a more "cosmic" form of his ideas [Trémaux 1874], which are reminiscent of Spencer's cosmic evolution.

WALLACE'S AND WEISMANN'S ADAPTATIONIST DEFINITION

Alfred Russel Wallace, codiscoverer of natural selection as an agent of evolution with Darwin, never really admitted the action of anything else in evolution. It followed, therefore, that he would insist that natural selection was the agent of speciation and hence that species are to be identified with their special adaptations.

Before he had gone public with his own evolutionism, Wallace asked if there was only an indefinable amount of difference that separated permanent varieties from species:

> If there is no other character, that fact is one of the strongest arguments against the independent creation of species, for why should a special act of creation be required to call into existence an organism differing only in degree from another which has been produced by existing laws? If an amount of permanent difference, represented by any number up to 10, may be produced by the ordinary course of nature, it is surely most illogical to suppose, and very hard to believe, that an amount of difference represented by 11 required a special act to call it into existence. [Wallace 1858, quoted in Kottler 1978: 294]

Kottler describes how Wallace's idea of species in this period involved lack of interbreeding, like Darwin's: "contact without intermixture being a good test of specific difference" [Kottler 1978: 295]. In his *Darwinism* [Wallace 1889: 167], he defined species as "[a]n assemblage of individuals which have become somewhat modified in structure, form, and constitution, so as to adapt them to slightly different conditions of life; which can be differentiated from allied assemblages; which reproduce their like; which usually breed together; and, perhaps, when crossed with their near allies, always produce offspring which are more or less sterile *inter se*" [cf. Romanes 1895, volume 2, p. 236].

Earlier, in his *Contributions to the Theory of Natural Selection* [Wallace 1870: 142], while discussing the numbers of species in the Malayan Archipelago of Papilionidae (a group of butterflies), he discussed the principle by which he should give them the specific rank:

> One of the best and most orthodox definitions is that of Pritchard [sic],[9] the great ethnologist, who says, that "separate origin and distinctness of race, evinced by a constant transmission of some characteristic peculiarity of organization" constitutes a species. Now leaving out the question of "origin" which we cannot determine, and taking only the proof of separate origin, "the constant transmission of some characteristic peculiarity of organization," we have a definition which will compel us to neglect altogether the amount of difference between any two forms, and to consider only whether the differences that present themselves are permanent. The rule, therefore, I have endeavoured to adopt is, that when the difference between two forms inhabiting separate areas seems quite constant, when it can be defined in words, and when it is not confined to a single peculiarity only, I have considered such forms to be species.

This is a generative conception again. He then discusses variation, polymorphisms, local varieties, coexisting varieties, and subspecies, saying:

> Species are merely those strongly marked races or local forms which when in contact do not intermix, and when inhabiting distinct areas are generally believed to have had a separate origin, and to be incapable of producing a fertile hybrid offspring. . . . [I]t will be evident that we have no means whatever of distinguishing so-called "true species" from the several modes of variation here pointed out, and into which they so often pass by an insensible gradation. [p. 161]

Wallace is taking Darwin's approach to its natural conclusion, as it were. If varieties are incipient species, and species merely well-marked and more or less permanent varieties, then there is nothing *sui generis* about *species* as a categorical rank. In this he was followed also by August Weismann [1904], who treated species entirely as complexes of adaptations. Discussing a number of examples of series of forms that can only be arbitrarily delimited, Weismann notes:

> All the individual members of these series are connected by intermediate forms in such a manner that a long period of constancy of forms seems to be succeeded by a shorter period of transformation, from which again a relatively constant form arises.

> We see, therefore, that the idea of species is fully justified in a certain sense; we find indeed at certain times a breaking up of the fixed specific type, the species becomes variable, but soon the medley of forms clears up again, and a new constant form arises—*a new species*, which remains the same for a long series of generations, until ultimately it too begins to waver, and is transformed once more But if we were to place side by side the cross-sections of this genealogical tree at different levels, we should only see several well-defined species between which no intermediate forms could be recognized; these would only be found in the intermediate strata. [Volume II, p. 305]

> [T]he species is essentially a complex of adaptations, of modern adaptations which have been recently acquired, and of inherited adaptations handed down from long ago—a complex which might well have been other than it is, and indeed must have been different if it had originated under the influence of other conditions of life. [p. 307]

In contradiction to Nägeli, who thinks of species as "a vital crystallization" [in Weismann's words, p. 307], Weismann denies that there is an evolutionary force that impels species to evolve, and defends natural selection as the entirety of evolutionary mechanism. Since the major features of evolutionary groups are adaptive in their origin, "if the step from one species to the next succeeding one does not depend on adaptation, then the greater steps to genera, families, and orders cannot be referred to it either, since these can only be thought of as depending upon a long-continued splitting up of species" [p. 306].

Weismann appears to think that entire species transmute and are changed into new species after a period of fragmentation of forms. However, he realizes that species are variable: "But of course species are not exclusively complicated systems of adaptations, for they are at the same time 'variation complexes,' the individual components of which are not all adaptive, since they do not all reach the limits of the useful or the injurious" [p. 307]. He recognizes that selection applies to suborganismic "vital units" [p. 308] and that there are "indifferent characters," or non-adaptive characters, which result as by-products, as it were, of selection for "a harmonious whole." Selection in the "germ plasm" may "give rise to correlative variations in determinants next to them or related to them in any way, and that these may possess the same stability as the primary variation. This seems to me sufficient reason why biologically unimportant characters may become constant characters of the species" [p. 308]. An example of this is the vestigial hind limbs in the Greenland whale [p. 313]. So Weismann was not exactly the panadaptationist he is sometimes made out to be [for example, by Gould 2002: 198ff.], and he allowed for the existence of neutral characters. However, he rejected outright the views of those who thought that isolation was a precondition to new species and that the characters that formed them were in any way neutral. In discussing variation, he notes [p. 286] that

> there are very variable species and very constant species, and it is obvious that colonies which are founded by a very variable species can hardly ever remain exactly identical with the ancestral species; and that several of them will turn out differently, even granting that the conditions of life be exactly the same, for no colony will contain all the variants of the species in the same proportion, but at most only a few of them, and the result of mingling these must ultimately result in the development of a somewhat different form in each colonial area.

This is an early forerunner of the "founder effect" conception of the origin of new species proposed by Mayr [1954] and developed further by Hampton Carson [Carson et al. 1970; Carson 1971, 1975; Coyne 1994]. What is most striking about this is that Weismann is effectively ascribing speciation in this case to stochastic sampling. This is something that, as the strict selectionist Romanes held him to be (see the next chapter), he should not have adopted. Weismann opposed, though, an exclusivist position like that of Wagner's that *all* species had to be formed in this way.

THE SPECIES PROBLEM ARISES

OTHER DARWINIANS: E. RAY LANKESTER, GEORGE ROMANES, T. H. HUXLEY, E. B. POULTON, AND KARL JORDAN

At least one of Darwin's most prominent followers, E. Ray Lankester, exceeded Darwin's published suggestion that the term *species* was arbitrary. Poulton said that Lankester was "inclined to think that we should discard the word species not merely momentarily but altogether" [Poulton 1903: 62]. Ernst Haeckel concurred, stating in his *The Evolution of Man* [1874, third edition cited in Haeckel 1896, volume 1, p. 115] that

> [e]ndless disputes arose among the "pure systematizers" on the empty question, whether the form called a species was "a good or bad species, a species or a variety, a sub-species or a group," without the question being even put as to what these terms really contained and comprised. If they had earnestly endeavoured to gain a clear conception of the terms, they would long ago have perceived that they have no absolute meaning, but are merely stages in the classification, or systematic categories, and of relative importance only.

T. H. Huxley, who had disagreed in correspondence with Darwin over saltative evolution, likewise wrote of species [Huxley 1906: 226f.] that

> [a]nimals and plants are divided into groups, which become gradually smaller, beginning with a KINGDOM, which is divided into SUB-KINGDOMS; then come the smaller divisions called PROVINCES; and so on from a

PROVINCE to a CLASS, from a CLASS to an ORDER, from ORDERS to FAMILIES, and from these to GENERA, until at length we come to the smallest groups of animals which can be defined one from the other by constant characters, which are not sexual; and these are what naturalists call SPECIES in practice, whatever they may do in theory.

If in a state of nature you find any two groups of living beings, which are separated one from the other by some constantly-recurring characteristic, I don't care how slight or trivial, so long as it is defined and constant, and does not depend on sexual peculiarities, then all naturalists agree in calling them two species; that is what is meant by the word species—that is to say, it is, for the practical naturalist, a mere question of structural differences.[1]

[1] I lay stress here on the *practical* signification of "Species." Whether a physiological test between species exist or not, it is hardly ever applicable by the practical naturalist.

He repeats requirement for the use of nonsexual characters again in a later essay [pp. 301ff.]. Huxley treats species in this limited taxonomic context[1] as merely conventional definitional entities—that is, as a nominal essence. The note implies that he expects there may be some real essence in the form of physiological (i.e., reproductive?) differences but that they are not useable by practicing taxonomists. So long as there is a constant unvarying character and a smallest diagnosable group, there is a species. In his 1859 review of the *Origin*, when discussing Darwin's views on sterility of hybrids, Huxley asks what is known of the "essential properties of species" [Huxley 1893a: 50]. He summarizes: "Living beings, whether animals or plants, are divisible into multitudes of distinctly defineable kinds, which are morphological species. They are also divisible into groups of individuals, which breed freely together, tending to reproduce their like, and are physiological species."

However, Huxley, following and in concert with Darwin, denied that reproductive isolation was sufficient to make a species and noted that the degree of infertility between species varies from absolute to minor [p. 50]. Elsewhere, in his 1863 critical review of the *Origin* [reprinted in Huxley 1906 as Chapter IX], he distinguishes between internal ("physiological") and external ("anatomical" or "morphological") characterizations of species [Forsdyke 2001: 31], and in a paper entitled "Darwin on the Origin of Species" published in the *Westminster Review* of 1860 [Chapter XIII], he notes that naturalists employ the term *species* in a double sense to denote "two very different orders of relations":

When we call a group of animals, or of plants, a species, we may imply thereby either, that all these animals and plants have some common peculiarity of form or structure; or, we may mean that they possess some common functional character. That part of biological science which deals with form and structure is called Morphology—that which concerns itself with function, Physiology—so that we may conveniently speak of these two senses or aspects of "species"—the one as morphological, the other as physiological. [p. 302]

George Romanes, who coined the term *neo-Darwinism* to sneer at the extreme selectionist views of Wallace and Weismann,[2] in his *Darwin and after Darwin* gave five definitions of species [1895, volume 2, pp. 229–231]:

1. *A group of individuals descended by way of natural generation from an original and specially created type.* He calls this "virtually obsolete."

2. *A group of individuals which, while fully fertile inter se, are sterile with all other individuals—or, at any rate, do not generate fully fertile hybrids.* Romanes calls this the "physiological definition" and claims that it is not entertained by any naturalist at that time, as it is incomplete.

3. *A group of individuals which, however many characters they share with other individuals, agree in presenting one or more characters of a peculiar kind, with some certain degree of distinctness.* Romanes claims this is practically followed by all naturalists. But it is insufficient to enable a uniform standard of specific distinction, so he adduces two more definitions, "which will yield to evolutionists the steady and uniform criterion required."

4. *A group of individuals which, however many characters they share with other individuals, agree in presenting one or more characters of a peculiar and hereditary kind, with some certain degree of distinctness.* This merely adds hereditary characters to the previous definition.

And finally he adds one for the "ultra-Darwinians" who insist that species are formed through natural selection.

5. *A group of individuals which, however many characters they share with other individuals, agree in presenting one or more characters of a peculiar, hereditary, and adaptive kind, with some certain degree of distinctness.*

These are the "logically possible" definitions of species, meaning that they include all the differentiae of the thing defined. Romanes presents something rather similar to some versions of the modern phylogenetic species concepts, but the "degree of distinctness" is still relatively vague and relies on the judgment of the specialists. He rejects absolutely the idea that the characters that mark species from each other must or even can be adaptive, and a discussion he instigated in the Biological Section of the British Association ended in "as complete a destruction as was possible of the doctrine that all the distinctive characters of every species must necessarily be useful, vestigial or correlated. For it became un-questionable that the same generalization admitted of being made, with the same degree of effect, touching all the distinctive characters of every 'snark'" [p. 235].

Lewis Carroll must have been pleased. Romanes then takes Wallace to task for his definition, and Weismann for his rejection of the inheritance of acquired characters [p. 241], which Romanes as a "proper" Darwinian had followed Darwin in defending; in fact, this topic was the occasion for the "more Darwinian than Darwin" comment when he coined the terms *neo-Darwinian* and *ultra-Darwinism*. He distinguished between species formed through nonhereditary influence of the environment as *somato-genetic* species, and those in which the environment makes changes to hereditable material—the germ plasm, in Weismann's terms—as *blasto-genetic species*. Neither term survived the rejection of neo-Lamarckism with the rise of Mendelian genetics in 1900 [Bowler 1989b].

E. B. Poulton wrote an extensive and influential essay on species [Poulton 1903] in which he coined the term *syngamy*. He noted that the fixity of species was not required by Augustine or Aquinas (itself arguable) and cites Aubrey Moore, who fingers Milton as the guilty party for this doctrine, although he thinks it was due to the spirit of the age (which, as we have seen, is rather correct; the received view of the middle ages was nonhistorical). He also cites Sir William Thistleton-Dyer as claiming that fixity was traceable to Bauhin (1550–1624) and Jung (1587–1657), and he discusses Darwin's own views in some detail. He coined several terms to set up the discussion about species, some of which have entered into common usage [pp. 60–62]. First, he defines groups formed by Linnaean diagnosis of forms as *Syndiagnostic*. Then, he names groups that freely interbreed as *Syngamic*, and in a privative fashion terms those that do not as *Asyngamic* (with the substantive noun forms *Syngamy* and *Asyngamy*). Next, he coins the term *Epigony* to mean breeding from a common parent. Finally, he provides a term for the organic forms that live in the same region *Sympatric*. Again, he uses a privative term—*Asympatric*—for those that do not. *Allopatry* had to wait for Mayr.[3]

With the technical apparatus in hand, Poulton moves from diagnosis to the underlying reality of species. He says, "Diagnosis . . . is founded upon the conception that there is an unbroken transition in characters of the component individuals of a species. Underlying this idea are the more fundamental conceptions of species as groups of individuals related by Syngamy and Epigony" [p. 64]. He argues, contrary to Darwin's views, that sterility is an effect of *Asyngamy*, not vice versa; reiterating Max Muller's and Moritz Wagner's views on *speciation*, as the topic came to be known later. Thistleton-Dyer had said that older writers employed "the word species as a designation for the totality of all individuals differing from all others by marks or characters which experience showed to be reasonably constant or trustworthy, as is the practice of modern naturalists" [p. 66]. He (inaccurately, as noted earlier here) cites Darwin a week after the publication of the *Origin*: "I met Phillips, the palaeontologist, and he asked me, 'How do you define a species?' I answered, 'I cannot.' Whereupon he said, 'At last I have found out the only true definition,—any form which has ever had a species name!'" [Poulton cites *More Letters of Charles Darwin* from 1903 (Darwin 1972), Volume I, p. 127].[4] Similar views were later held by C. Tate Regan [1926] and were named by Blackwelder [1967] the "taxonomic species concept" and by Kitcher [1984] the "cynical species concept."

According to Poulton, though, species are interbreeding communities, syngamic communities. This came later to be termed, briefly [Lotsy 1931], a *Syngameon* as a neutral term for such communities, irrespective of their rank, although as Dobzhansky noted [1941: 311], Lotsy seemed to equate the syngameon with "species" anyway. It underlay the gradual variation in forms from one end of a range of organisms to another, which he referred to as a transition. Although, when clear, this approach made diagnosis possible, there were cases in which it would fail: polymorphisms, seasonal dimorphisms, individual developmental adaptation (he calls it *individual modification* and cites Baldwin), geographic races and subspecies, and artificial selection. Moreover, says Poulton, interspecific sterility is not an infallible test of specificity, because, as Darwin knew, related forms often can interbreed. Instead, sterility of hybrids is "an incidental consequence of asyngamy" [p. 80], of separation for a long period. Moreover, asyngamy itself is usually the by-product of asympatry [p. 84], although he allowed that Karl Jordan was correct when he said that it could be due to *mechanical incompatibility* [Jordan 1896; cf. later Jordan 1905a, 1905b], or the lack of fit between sexual organs (in the 1896 report using the bird genus *Papilio* as the test case). Poulton also accepts Henry Bates's 1862 claim of preferential mating. All this notwithstanding, sterility is not, in his opinion, due to the action of selection. This seminal paper, republished in 1905, was greatly influential in setting up the terms, both literally and metaphorically, of the twentieth-century debate over species and speciation. It subsequently inspired the writing of a text [Robson 1928] that summarized the issues as understood at the end of the period in which neo-Lamarckian mechanisms were still viable hypotheses, shortly before Dobzhansky's paper and book. We may usefully date the Species Problem from Poulton's paper.

Karl Jordan's papers [especially Jordan 1905b] seem to have had some impact on the way people discussed variation within species, and they contain the core of the biospecies concept later propounded by Mayr, who mentions Jordan as a forerunner [Mayr 1982: 272] and quotes Jordan's statement "Individuals connected by blood relationship form a single faunistic unit in an area . . . the units, of which the fauna of an area is composed, are separated from each other by gaps which at this point are not bridged by anything."[5]

Jordan gave the criteria that he believed defined two organisms as being in the same or different species [p. 159] in the context of a discussion of the variation of physical traits in organisms that we use to establish the "blood relationships of individuals and thus the physical gaps between species." He said that individuals of two species announce themselves by having physiological differences that they always reproduce in their progeny, and by being able to live in the same region without blending into each other. He listed three criteria: "The criteria of the *species* [he uses the Latin word *species* here—JSW] (= *Art*) concept are thus threefold, and each individual point is an applicable test: a species requires known traits, it does not beget descendents equally well with individuals of other species, and it does not coalesce into other species."[6]

Jordan's view seems to require not only geographic isolation but also constancy of characters, which Mayr's later definition does not. He goes on to say that the noncoalescence (non-hybridization) of species explains the "enormous number" of extant species and that this is due to the internal organization of the species (by which I take him to mean of the typical genetic structure of the species). He makes the comment that species act as if there were "*no* relationship between them, but as if they were doing business for themselves."[7]

NON-DARWINIAN IDEAS AFTER DARWIN

However, the variable ensemble of Darwinian ideas [Hull 1973a] was not universally adopted. During the so-called eclipse of Darwinism period [Bowler 1983], in which neo-Lamarckian ideas overtook Darwin's mechanism of natural selection, species were often thought to be types again. Such American neo-Lamarckians as Cope and Hyatt adopted an "orthogenetic Lamarckism" in which species underwent a series of developmental changes in a kind of embryological analogy between individuals and species [Bowler 1983: 121ff.]. In the period from Edward Drinker Cope's 1868 essay "On the Origin of Genera" [Cope 1868] through to the period immediately before World War I, species were thought by this school to be the result of internal forces rather than selection or geographic isolation. Cope called this growth force "bathmism" while others, such as Alpheus Hyatt and Alpheus Packard, and Henry Fairfield Osborn, had other mechanisms. Osborn, in particular, adopted the Baldwin Effect as (he thought) a non-Darwinian mechanism [Bowler 1983: 131] that enabled organisms to direct their own evolution through individual adaptation, which tarred the Baldwin Effect as anti-Darwinian or Lamarckian for a long time to come [Turney, Whitley, and Anderson 1996]. Osborn, under attack from Baldwin for saying that variation is non-random, then made the claim that there were linear variational trends for which Darwinism could not account [p. 132f.].

William Bateson [1894] produced a large book describing the sorts of variations that occurred from the type, in which he treated species as morphological classes, with no continuity of form between species and hence no reason to think they varied sufficiently for Darwinian evolution to occur. Species are groups of organisms united by a common form, but there is no definite difference that divides them. He writes of what he calls "the Problem of Species":

> No definition of a Specific Difference has been found, perhaps because these Differences are indefinite and hence not capable of definition. But the forms of living things, taken at a moment, do nevertheless most certainly form a discontinuous series and not a continuous series. . . .
>
> The existence, then, of Specific Differences is one of the characteristics of the forms of living things. This is no merely subjective conception, but an objective, tangible fact. This is the first part of the problem.
>
> In the next place, not only do Specific forms exist in Nature, but they exist in such a way as to fit the place in Nature in which they are placed; that is to say, the Specific form which an organism has, is *adapted* to the position which it fills. This again is a relative truth, for the adaptation is not absolute. [p. 2f.]

For Bateson, though, form is more than a way of describing or diagnosing species; it *is* what species are—they are classes of forms. Adaptation is not relevant to the origin of these discontinuous forms, but variation is. Bateson stops short of saying that form is a causal factor in evolution, but he does say that symmetry of form and the repetition of forms (*merism* is his term for this) are causes of speciation [pp. 19ff.]. Indeed, variation from the type (the Specific Differences) is due to a "pathological accident," as he approvingly quotes Virchow [p. 75]. Later, as one of the Mendelian geneticists, Bateson opposed the idea that there was genetic variation of the kind Darwinian selection required to form species. In 1913, he wrote:

> All constructive theories of evolution have been built upon the understanding that we know if the relation of varieties to species justifies the assumption that the one phenomenon is a *phase* of the other, and that each species arises . . . from another species either by one, or several, genetic steps. . . . [However,] complete fertility of the results of intercrossing [between members of different "species"] is, and I think must rightly be regarded, as *inconsistent* with actual specific difference." [quoted in Forsdyke 2001: 31]

Another Mendelian, Hugo de Vries, shortly after proposed a concept of "elementary species" as pure genetic lines in his 1904 lectures "Species and Varieties" [de Vries 1912] and in the earlier *Die Mutationstheorie* [de Vries 1901, 1911]. According to the Mendelian view that de Vries adopted, species in the Linnaean sense were actually comprised of a number of smaller lines of pure genetic stock:

> Species is a word, which has always had a double meaning. One is the systematic species, which is the unit of our system. But these units are by no means indivisible. . . . These minor entities are call varieties in systematic works. . . . Some of these varieties are in reality just as good as species, and have been "elevated," as it is called, by some writers, to this rank. This conception of the elementary species would be quite justifiable, and would get rid of all difficulties, were it not for one practical obstacle. The number of species in all genera would be doubled and tripled, and as these numbers are already cumbersome in many cases, the distinction of the native species of any given country would lose most of its charm and interest.
>
> In order to meet this difficulty we must recognize two sorts of species. The systematic species are the practical units of the systematists and florists, and all friends of wild nature should do their utmost to preserve them as Linnaeus has proposed them. These units, however, are not really existing entities; they have as little claim to be regarded as such as genera and families. The real units are the elementary species; their limits often apparently overlap and can only in rare cases be determined on the sole ground of field-observations. Pedigree-culture is the method required and

any form which remains constant and distinct from its allies in the garden is to be considered as an elementary species. [de Vries 1912: 11]

De Vries came to his views through the observation of what we now know to be alloploid forms in *Oenothera lamarckiana*, the evening primrose, which de Vries had cultivated and maintained pedigrees [p. 17, Lecture IX]. He was of the view that these were straight mutations forming a single new elementary species at once. In a way, given the way alloploids occur, he was right, but his idea led fairly directly to the later views of Goldschmidt, who, like de Vries, felt that species *always*, or almost always, arose in sudden saltative leaps.

In the lecture discussing the evening primrose, de Vries further defines the marks of an elementary species: "Elementary species differ from their nearest allies by progressive changes, that is by the acquisition of some new character. The derivative species has one unit more than the parent" [p. 253]. This meant that if the new elementary species were crossed with its parental elementary species, the progeny would be incomplete for that character (he clearly means Mendelian factor) and would therefore be unnatural [p. 254]. Hence, backcrossing would not occur in the wild or under cultivation [cf. p. 527]. Elementary species thus do not exhibit subvarieties for they are the "real type" [p. 127]. De Vries's concept was fundamentally anti-Darwinian, in the sense that he rejected the idea of there being continuous variation within species on which selection could act in such a way as to form new species. In fact, he argued for his theory on the grounds that the length of time required for evolution would be noticeably shorter on his account. At that time, Lord Kelvin's arguments against Darwinian evolution—that reasoning from the rate of cooling of the earth, evolution would need to have happened in tens of millions, not hundreds or thousands of millions, of years—were still current. Rayleigh's discovery of radioactivity as a source of planetary heat was not announced until about this time (1906), and it did not immediately filter through to the wider scientific community [see Bowler 1989a: 207].

Prior to Bateson's and Poulton's essays, there was no species *problem* as such but only a species *question*. The latter is concerned primarily with the origins of species, how they come to be. The *problem* arises when we have accounts of species formation, whether by selection or something else, that do not involve immediate saltation from one to another or *creatio de novo*. It is the problem of defining what rank it might be that species achieve when they become species, and this sets the agenda for Dobazhansky's paper (discussed later) and the remainder of the twentieth-century debates.

There were a few Darwinians in the period before the Synthesis. For example, in 1934, J. Arthur Thompson defined species extensively in the classical terms but with an emphasis on the role played by selection [Thompson 1934, volume 2, p. 1333f.]. Thompson defines it in the context of the human species and races of man and gives four criteria: nontrivial difference, true breeding and constancy of characters, interfertility and production of fertile offspring, and the constancy of specific characters in different environments. He says, "To sum up: *A species is a group of similar individuals differing from other groups in a number of more or less true-breeding characters, greater than those which often occur within the limits of a family, and not the direct result of environmental or other nurtural influences. The members of a species are fertile with one another, but not readily with other species*" [p. 1334, italics in original]. Races share some of these features (that is, they breed true), but they are less marked than specific characters, and they may be maintained by natural and sexual selection. However, Darwinian notions of species were the exception rather than the rule for some time into the new century. Of more significance was the influence of Mendelian genetics and, importantly, hybridization

JOHANNES PAULUS LOTSY AND THE EVOLUTION OF SPECIES BY HYBRIDIZATION

In a work published in English [Lotsy 1916], Dutch botanist Johannes Paulus Lotsy (1867–1931) proposed both a definition of species and a conception of evolution. He begins by noting that the concept of species is vague: "All theories of evolution have, until quite recently, been guided by a *vague* knowledge of what a species is, and consequently have been vague themselves" [p. 14]. Lotsy therefore proposes a definition based on "identity of constitution" and, citing Ray, discusses Jordan's discovery of variety within all Linnaean species. He says, "Jordan *consequently discarded morphological comparison as a criterium for specific purity* and, falling back to Ray (whom he may or may not have known) *substituted for it: nulla certior . . . quam distincta propagatio ex semine*" [p. 21, italics in original].

From this, he says, Jordan drew the "well founded conclusion": "*The Linnaean species is no species*" [p. 22]. Lotsy therefore proposed a term, the *Linneon*, for the product of Linnaean classification defined as "*the total of individuals which resemble one another more than they do any other individuals*" (italics in original). The types contained *within* a Linneon Jordan called species, and Lotsy calls *Jordanons*, since "*[b]reeding true to type is . . . by itself no reliable test for specific purity*" [p. 23]. He then gives his own, proper definition of a species: "*A species consists of the total of individuals of identical constitution unable to form more than one kind of gametes*" [p. 23]. Moreover, Lotsy proposes a genetic test: "*Specific purity is indicated by the uniformity and identity of the F1 generations obtained by crossing the individuals to be tested, RECIPROCALLY*" [p. 24].

Thus, a species is for Lotsy an operationally applicable concept. The result is three definitions of terms [p. 27]:

LINNEON: *to replace the term species in the Linnaean sense, and to designate a group of individuals which resemble one another more than they do any other individuals.*

To establish a Linneon consequently requires careful morphological comparison only.

JORDANON: *to replace the term species in the Jordanian sense, viz: mikrospecies* [sic], *elementary species etc. and to designate a group of externally alike individuals which all propagate their kind faithfully, under conditions excluding contamination by crossing with individuals belonging to other groups, as far as these external characters are concerned, with the only exception of noninheritable modifications of these characters, caused by the influences of the surroundings in the widest sense, to which these individuals or those composing the progeny may be exposed.*

To establish a Jordanon, morphological comparison alone consequently does not suffice; the transmittability of the characters by which the form was distinguished, must be experimental breeders.

SPECIES: *to designate a group of individuals of identical constitution, unable to form more than one kind of gametes; all monogametic individuals of identical constitution consequently belonging to one species.*

Lotsy rejects the idea of intraspecific variation in Darwin's sense, and he is of the view that every homozygotic form is itself a species. It follows that every mixing of these "pure forms" is the origination of a new species if that novelty persists. Moreover, he thinks that species can be polyphyletic—they can arise more than once, because a species is formed in virtue of its "constitution," not in terms of its history [p. 45]. Moreover, "nature primarily can make nothing but individuals" [p. 46], and it can secondarily group those individuals in various ways. Linneons are not natural, though—they are groupings formed by the human mind. He gives an example using human races [pp. 47–49], supposing that even if there were four "pure races" arranged in army battalions, we could divide them up in various ways, according to tattoo marks given from parent to child so that each child has two marks. If there has been no intercrossing, then although the individuals have changed, he says, the constitution would remain the same, and we could arrange the progeny into those groups; but if crossings occurred, we could not assign them to the right armies.

In the forefront of his definitions, Lotsy has the Mendelian genetics then being first investigated in detail. He thinks that classification by the genetic constitution forms a kind of Lockean "real essence," which is the point of the race example. If our groupings match nature's real essence (that is, genetic constitution), then they are natural groups. Otherwise they are not, and given the typological nature of his definition, ordinary species (i.e., what came to be called "biological species") were not natural entities. The remainder of his argument relies on Mendelian assortment forming novel varieties, as we would call them, or "allogamous forms" as he calls them [p. 159], from mutations.

GÖTE TURESSON: ECOSPECIES AND AGAMOSPECIES

The Swedish botanist Göte Turesson undertook a series of experiments in the 1920s and early 1930s [Turesson 1922a, 1925, 1927, 1930], in which he transplanted the "ground stock" of widely distributed Swedish plants into various different habitats—dunes, sea cliffs, woodlands, high altitudes, and so on [Turrill 1940: 52]—and discovered that differing forms arose as a response to climate. He made a number of influential distinctions on taxa that later formed the basis for ecological species concepts [Turesson 1922b, 1929]. He proposed *ecospecies* "to cover the Linnean species or genotype compounds as they are realized in nature" and related *coenospecies* (the Linnaean taxon) to ecospecies in a diagram (figure 7 [Turesson 1922a: 344]).

Coenospecies are "the total sum of possible combinations in a genotype compound," and include one or more ecospecies, which include one or more *ecotypes*, the forms that develop in different *ecosystems* or habitat types. These are comprised of all the "reaction-types" of ecotypes that are elicited by extreme habitats, called *ecophenes*. In a similar inclusive hierarchy, he listed a genetical array of concepts—*genospecies* (the genetical construction of ecospecies), *genotypes* (Johannsen's 1909 term),

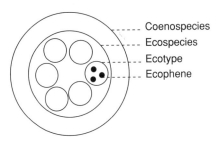

Figure 7 Turesson's view of coenospecies and ecospecies. Redrawn from Turesson [1922b: 344].

and the "reaction-types" of genotypes, *genophenes*. This dual hierarchy between the ecological and genealogical was repeated later by Eldredge [1989] and Salthe [1985]. Turesson's "reaction-types" are in modern terms the reaction norms of genes, although the idea of a reaction norm for an ecotype appears to have been abandoned.

In one paper [Turesson 1929: 332–333], he gave succinct definitions of the major terms:

a) *Ecospecies*: An amphimict-population the constituents of which in nature produce vital and fertile descendents with each other giving rise to less vital or more or less sterile descendants in nature, however, when crossed with constituents of any other population. . . .

b) *Agamospecies*: An apomict-population the constituents of which, for morphological, cytological or other reasons, are to be considered as having a common origin. . . .

c) *Coenospecies*: A population-complex the constituents of which group themselves in nature in species units of lower magnitude on account of vitality and sterility limits having all, however, a common origin so far as morphological, cytological or experimental facts indicate such and origin.

Turesson criticizes the Linnaean conception of species for not helping us to determine the *natural* limitations of species. He says that the Linnaean notion covers several senses, defined earlier, and that the "existence in nature of units of these different orders also makes it a logical impossibility to reach one standard definition of the 'species'" [p. 332]. This is, incidentally, the first use of the term "agamospecies" I have encountered, so he may very well have coined it. As far as I can tell, Turesson does not require that agamospecies cannot also be ecospecies or coenospecies.

Mayr [1982: 277] accuses Turesson of being typological, and claims that he gives the impression of plant species being comprised of "a mosaic of ecotypes rather than as an aggregate of variable populations," but this is not how contributors to the 1940 *New Systematics* volume read him. Turrill [1940: 52] notes, "He [Turesson] has clearly shown not only that the species, as usually accepted by the taxonomist, is a complex assemblage of biotypes, but also that the species population varies in its biotype composition with habitat conditions." Another contributor [Salisbury 1940: 332] also regards Turesson's approach as a populational one. In any event, it is clear that Turesson's conception of species taxa included *both* a "biological" (i.e., a genetic) and an ecological aspect. It may be that Turesson's scheme is itself typological, but its implementation and later influence are not, in themselves, necessarily so.

GERMAN THINKERS: ISOLATION AS THE KEY

Species concepts played an increasing role in the thinking of several German-speaking biologists in the early part of the twentieth century. In particular, the views of Erwin Stresemann, curator of birds at the Berlin Museum and mentor to both Ernst Mayr and Bernhard Rensch, were influential. In 1919, he had written that morphology as a criterion of species had been abandoned by the late 1890s among ornithologists, in favor of physiological divergence, as evidenced by reproductive isolation. He said [Stresemann 1919: 64, quoted in Mayr 1980: 415], "[F]orms of the rank of species have physiologically diverged from each other to such an extent, that they can come together again without mixing with each other." "Morphological divergence is thus independent of physiological divergence" [Stresemann 1919: 66].

Bernhard Rensch was strongly influenced by this approach also [Rensch 1980: 294f.], and he defined several terms to deal with the reproductive isolation of species and within species in the case of races that do not interbreed, although overall the species has a shared gene pool. He called these *Rassenkreise*, or "race circles"; and complexes of incipient species that replaced each other geographically, he called *Artenkreise*, or "species circles" [Rensch 1928, 1929; see also Rensch 1947, 1959]. Although he doesn't define species in the latter work, he does talk frequently about "good species" being those that are isolated by sexual or genetic differences when in contact.

A review of the state of play at the end of the period before the synthesis began was published by zoologist Guy C. Robson [1928], and it is of interest how much of the modern debate was prefigured there—polymorphisms, reproductive isolation, allopatry (under another name, of course), and genetic variance are all discussed. At that time, however, it was unclear whether or not the neo-Lamarckian view of the inheritance of acquired characters was a viable view or not. Robson tended to think not, but he allowed that later research may show otherwise. For him, species are not the necessary outcome of evolution and are recognized by correlations of differentiated characters [p. 223] that "hang together" [p. 224]. He notes that sampling of specimens fails to reproduce the wider diversity of characters of the larger assemblage of the taxon. Races are localized more or less homogeneous groups in what is a continuous distribution of forms. He rejects inability or disinclination to hybridize as a good test of species and also the proposal to give up the notion of species [p. 21]. Although there are no absolute criteria for judging species, Robson does think that well-defined groups do occur in nature [p. 22], and he rejects the "morphological," "genetic," "physiological," and "ecological" criteria as being either incomplete or ill-defined tests of species rank.

THE MENDELIANS: THOMAS HUNT MORGAN AND ALFRED H. STURTEVANT

Famous Chicago geneticist Thomas Hunt Morgan was initially of the view that species were nonexistent, arbitrary units devised for the convenience of taxonomists (and hence is a conventionalist species denier at this stage). He held that, with Buffon, only individual organisms existed: "We should always keep in mind the fact that the individual is the only reality with which we have to deal, and that the arrangement of these into species, genera, families, etc. is only a scheme invented by man for purposes of classification. Thus, there is no such thing in nature as a species, except as a concept of forms more or less alike" [Morgan 1903, quoted in Allen 1980: 359f.]. His view of species as types determined by convention never changed thereafter, although he did later indicate that we think of species in terms of their adaptations. He did allow that there was an enormous amount of variability in genes, though.

Morgan's students, in particular Alfred H. Sturtevant, had a stronger interest in species. Sturtevant held that there was a "wild type" of each species from which variants were insignificant and that species actually differed in few genes [Dobzhansky 1980]. H. J. Muller, however, later adopted in full the neo-Darwinian account of the modern synthesis

[Muller 1940], treating species as formed through hybrid infertility due to genetic variations such as chromosomal rearrangements. Even in his *Drosophila*, the nature of species was plastic and variable [p. 252f.], and "[i]t becomes, then, a matter of definition and of convenience, in any given series of cases, just where we decide to draw the line above which two groups will be distinguished as separate species, and below which they are denoted subspecies or races, since in nature there is no abrupt transition here" [p. 253]. Despite this, he said, a "well-knit" species is qualitatively different from individuals of two closely related species, and the word *species* denotes, to a "rough approximation, how these groups stand in relation to that general level at which 'speciation' takes place" [p. 254].

THE SYNTHESIS AND SPECIES

On balance, the abandonment of the term 'species' as a
category to express the findings of experimental taxonomy,
reflecting a clean break with the non-evolutionary past, would
seem to be the right policy. The reluctance of geneticists and
others to take this step is, perhaps, bound up with a lingering
belief that species are, in some way, 'real entities', not 'artificial
constructs of the human mind' as are other taxa. This belief
makes it unthinkable to abandon the term for such an important branch of biology as the study of micro-evolution. If, however, 'real entities' and 'artificial constructs' are regarded, not as
mutually exclusive, but as alternative descriptions, each valid in
its own context, this difficulty disappears.

Gilmour [1958]

RONALD FISHER AND WILD-TYPE SPECIES

Ronald A. Fisher is famous as the founder of the modern synthesis
between Mendelian genetics and Darwinian natural selection. *The
Genetical Theory of Natural Selection* [Fisher 1930] is a seminal work
that introduced mathematical models to genetics and selection, and while
often cited, it is rarely quoted. But Fisher addressed a number of questions in that book, including a rarely mentioned discussion about eugenics
(Fisher was in favor of a form of eugenics, and chapters 8 through 12
are an argument for it), and one of these, almost as a parenthetical comment, is about species in the context of sexual and asexual reproduction.

The tradition in British evolutionary biology since Fisher has, on the
whole, tended to treat species as names of convenience for communication, in the style of Darwin of the *Origin* and of Locke. It is therefore
somewhat surprising to note that Fisher had a realist approach to
species, one that predated the Dobzhansky 1935 article that is widely

seen as kicking off the species debate of the modern era (discussed later). Fisher says of species that "[t]he genetical identity in the majority of loci, which underlies the genetic variability presented by most species, seems to supply the systematist with the true basis of his concepts of specific identity or diversity" [p. 138]. Sexual species are, in fact, the "wild type" [p. 137]—the sum of all the genes in any species the great majority of which are uniform: "we have some reason to suppose that they [allelomorphic loci, or alleles in modern parlance—JSW] form a very small minority of all the loci, and that the great majority exhibit, within the species, substantially that complete uniformity, which has been shown to be necessary, if full advantage is to be taken of the chances of favourable mutations."

Fisher is here dealing with the existence of *asexual* species, as they present a problem for him, or rather, they would have if he had been (at that time) certain that any organisms existed without any sexual reproduction (the claim was not revised in the 1958 edition). In this case, he says:

> In such an asexual group, systematic classification would not be impossible, for groups of related forms would exist which had arisen by divergence from a common ancestor. Species, properly speaking, we could scarcely be expected to find, for each individual genotype would have an equal right to be regarded as specifically distinct. And no natural groups would exist bound together by constant interchange of their germ-plasm. [p. 135]

Clearly, this exchange of genes in germ-plasm is the *sine qua non* of a species for Fisher. But, he goes on to say, there *would* be an analogue of species in asexuals: "The groups most nearly corresponding to species would be those adapted to fill so similar a place in nature that any one individual could replace another, or more explicitly that an evolutionary improvement in any one individual threatens the existence of all the others."

So while Fisher is a realist about asexual groups, they are ecological groups adapted to the environment in which they find themselves and are kept distinct by virtue of selection against less fit variants. This resolves also the problem of favorable mutations—if a novelty of value arises in an asexual lineage, then it will not spread throughout the population but instead will replace the population. In sexual organisms that have proper species, selection maintains the identity of populations rather than of lineages, and favorable mutations can be spread by recombination of genes. But Fisher does not think that there will be many of these groups and that if they did exist, they would be those groups "of so simple a character that their genetic constitution consisted of a single gene" [p. 137].

Of sexual species proper, Fisher presents the view that apart from geographic isolation, in which "the two separated moieties thereafter evolv[e] as separate species, in almost complete independence, in somewhat different habitats, until such time as the morphological differences between them entitle them to 'specific rank'" [p. 139].

Species are also caused to fission by what we now call sympatric speciation, because in "many cases it may safely be asserted that no geographic isolation at all can be postulated." A species subject to different conditions at the extremes of its range will adapt at those extremes, and hybrid forms will be disadvantageous if the migration rate is less than the rate of increase of the favorable forms in the environment to which they are adapted. Fisher champions as a mode of speciation the selectionist account of Darwin and Wallace, yet he still allows for the Wagner-style mode of allopatric isolation.

At this point we have reached the beginnings of the modern synthesis and hence the modern debate. We are now equipped to put the modern debate into context, especially claims of conceptual novelty. To that topic we now turn, beginning with Theodosius Dobzhansky's discussion shortly after Fisher's book, which defined the modern species debate.

We may arbitrarily mark the beginnings of the modern debate with Dobzhansky's classic 1935 essay, "A Critique of the Species Concept in Biology," although Poulton's essay is also crucial. It's not really so arbitrary—from Dobzhansky's essay and the book that followed it [Dobzhansky 1937a] flowed both the present debate over what species are, and the birth of the modern synthesis. Dobzhansky's work was an attempt to take Darwin seriously about species by a working systematist not imbued with the British deflationary tradition. And he introduced two of the major innovations in the debate: the idea of species as evolutionary players, and the idea that genetic exchange marked out these players. It seems that when typostrophic views of evolution were abandoned, the issue from the Great Chain of what divisions nature forced on us and what divisions were of our own convenience came back to the fore. Later, Mayr described the synthesis as being a "shared species problem" [Mayr and Provine 1980]. It remains a shared problem.

THEODOSIUS DOBZHANSKY'S DEFINITION

Theodosius Dobzhansky was perhaps the most significant of all the synthesists[1] through the middle of the twentieth century. He introduced Sewall Wright's ideas on drift into the synthetic orthodoxy (often to considerable opposition), and his work on the laboratory and field genetics of *Drosophila* species revolutionized the field [cf. Depew and Weber 1995: 291–297, 300–302].

Dobzhansky published a paper in 1935, later substantially included in the chapter on species in his *Genetics and the Origin of Species* [Dobzhansky 1941], which discussed Lotsy's revision of Poulton's notion of syngamy as the foundation for a genetic population: "an habitually interbreeding community of individuals" [Dobzhansky 1941: 311]. Dobzhansky says of the syngameon approach to species that it applies only to panmictic populations of organisms and that, although attractive in its simplicity, it is therefore inapplicable in the case of many species that are divided into reproductively separated populations. He proposes instead to base specific rank on the existence of reproductive isolating mechanisms, and provides a revision to Lotsy's definition: "a species is a group of individuals fully fertile inter se, but barred from interbreeding with other similar groups by its physiological properties (producing either incompatibility of parents, or sterility of the hybrids, or both)" [Dobzhansky 1935: 353; cf. also Dobzhansky 1941: 312].

The *rank* of *species* is something that arises in evolution when continuity of reproduction becomes discontinuous: "Considered dynamically, the species represents that stage of evolutionary divergence, at which the once actually or potentially interbreeding array of forms becomes segregated into two or more separate arrays which are physiologically incapable of interbreeding" [Dobzhansky 1935: 354]. By "array of forms," Dobzhansky can be interpreted as meaning either diagnostic morphs, as Mayr does (discussed later), or as the types within a population that affect reproductive isolation. His discussion makes it clear that he does not intend "form" in an essentialistic sense, I believe, but in a causal sense. In the 1935 paper, he treats reproductive isolation in terms of physiological differences; this is the sense I interpret him to mean by "form," although he may have equivocated on the distinction between diagnostic and causal structures, as many did before and after him.

He had noted in *Genetics and the Origin of Species* the "taxonomic" definition [of an "affable taxonomist"—C. Tate Regan—that species are what competent systematists consider to be a species, p. 310], and he puts this failure to deliver a universal definition "that would make it possible to decide in any given case whether two given complexes of forms are already separate species or are still only races of a single species" down to the general method of species formation, "through a slow process of accumulation of genetic changes of the type of gene mutations and chromosomal reconstructions. This premise being granted, it follows that instances must be found in nature when two or more races have become so distinct as to approach, but not to attain completely, the species rank. The decision of a systematist in such instances can not but be an arbitrary one" [p. 310f.].

In short, evolution makes it impossible to determine if species rank has been reached. Nevertheless, the rank itself is real enough—it is the attainment of complete separation. He also notes that

> [w]e find aggregations of numerous more or less clearly distinct biotypes, each of which is constant and reproduces its like if allowed to breed. These constant biotypes are sometimes called elementary species, but they are not united into integrated groups that are known as species in the cross-fertilizing [i.e., sexual—JSW] forms. The term "elementary species" is therefore misleading and should be discarded. [Dobzhansky 1941: 320f.]

It seems Dobzhansky was influential in this regard, because de Vries's term did largely disappear from the debate after this. He also notes that biotypes can be cross-specific, and even cross-generic, and that

> [w]hich one of these ranks is ascribed to a given cluster is, however, decided by considerations of convenience, and the decision is in this sense purely arbitrary. In other words, the species as a category which is more fixed, and therefore less arbitrary than the rest, is lacking in asexual and obligatorily self-fertilizing organisms. . . .
>
> The binomial system of nomenclature, which is applied universally to all living beings, has forced systematists to describe "species" in the sexual as well as in the asexual organisms. Two centuries have rooted this habit so firmly that any thorough reform will meet with determined opposition. Nevertheless, systematists have come to the conclusion that sexual species and "asexual species" must be distinguished. . . . In the opinion of the writer, all that is saved by this method is the word "species." A realization of the fundamental difference between the two kinds of "species" can make the species concept methodologically more valuable than it has been.

In the 1951 edition [p. 275], Dobzhansky replaces the final sentence with "As pointed out by Babcock and Stebbins . . . , 'The species, in the case of a sexual group, is an actuality as well as a human concept; in an agamic complex it ceases to be an actuality.'" He also begins in that edition to discuss issues of typological thinking, under the influence of Mayr, in the chapter on populations, races, and subspecies, where he notes [pp. 268f.]:

> The classical race concept [of human races—JSW] was typological. . . .
> Typology is at the bottom of the vulgar notion that any so-called Negro in the United States . . . has a basic and unremediable Negroid nature, just as any Jew partakes of some Jewishness, etc. There are no Platonic types of Negroidness or Jewishness or of every race of squirrel or butterflies. Individuals are not mere reflections of their racial types; individual differences are the fundamental biological realities.

There is little hint of discussions of typology or essentialism in the earlier works. Dobzhansky deals extensively with variation within populations—it is the raison d'être of the book, and so it might be a case of the wood not needing to be specified when the trees are so well described. By 1970, though, he is well in line with the Mayrian program.

AFTER DOBZHANSKY: THE BEGINNINGS OF
THE MODERN DEBATE

At the time of the modern synthesis, announced in Julian Huxley's book by that title in 1942, there was little dispute among those involved that species were real enough, but there was a wide range of opinion about what that meant. Darlington [1940], for example, explicitly appealed to Ray's dictum (in Latin) that to sort living beings into species we need no more than "*distincta propagatio ex semine*" [p. 137] but that "[t]here are many kinds of species and many kinds of discontinuities between species" [p. 158]. He noted:

> We feel we ought to have a 'species concept'. In fact there can be no species concept based on the species of descriptive convenience that will not ensnare its own author so soon as he steps outside the group from which he made the concept. The only valid principles are those that we can derive, not from fixed classes but from changing processes. To do this we must go beyond the species to find out what it is made of. We must proceed (by collaboration) to examine its chromosomal structure and system of reproduction in relation to its range of variation and ecological character. From them we can determine what is the genetic species of Ray, the unit of reproduction, a unit which cannot be used for summary diagnosis, but which can be used for discovering and relating the processes of variation and the principles of evolution. [p. 159]

In contrast, Julian Huxley, in his introduction to *The New Systematics* [Huxley 1940: 16ff.], argues that Dobzhansky's 1937 definition "goes far beyond the facts." He notes the constancy of cross-fertilization among plants in particular and says that therefore "Dobzhansky's definition is untrue, or, if true, taxonomic practice must be so re-cast as to rob the term species of its previous meaning" [p. 17]. H. J. Muller's contribution to that volume [Muller 1940] agrees—there is no fixed rank dividing species from varieties or races [p. 258], although

> divergence goes on very differently, and much more freely, between those which can and do cross, and it is therefore justifiable and useful, even though difficult, to make the species distinction, if it is made in such a way as to correspond so far as possible with this stage of separation. At the same

time it must be recognized that the species are in flux, and that an adequate understanding of their relationships can be arrived at only on the basis of an understanding of the relationships between the minor groups and even between the individuals, supplemented by the study of the differences found through observations on the systematics of the larger groups.

In the book announcing the synthesis, Huxley later criticized Dobzhansky for underplaying the difficulties that a simple intersterility criterion encountered, particularly in plants.

The dynamic point of view is an improvement, as is the substitution of incapacity to exchange genes for the narrower criterion of infertility: but even so, this definition cannot hold, for it still employs the lack of inter-breeding as its sole criterion. "Interbreeding without appreciable loss of fertility" would apply to the great majority of animals, but not to numer-ous plants. In plants there are many cases of very distinct forms hybridizing quite competently even in the field. To deny many of these forms specific rank just because they can interbreed is to force nature into a human definition, instead of adjusting your definition to the facts of nature. Such forms are often markedly distinct morphologically and do maintain themselves as discontinuous groups in nature. If they are not to be called species, then species in plants must be deemed to differ from species in animals in every characteristic save sterility. [Huxley 1942: 162f.]

He gives his own criteria a few pages later, after determining that single-criterion definitions are useless [p. 164f.]:

In general, it is becoming clear that we must use a combination of several criteria in defining species. Some of these are of limiting nature. For instance, infertility between groups of obviously distinct mean type is a proof that they are distinct species, although once more the converse is not true.

Thus in most cases a group can be distinguished as a species on the basis of the following points jointly: (i) a geographical area consonant with a single origin; (ii) a certain degree of constant morphological and presum-ably genetic difference from related groups; (iii) absence of intergradation with related groups. . . . Our third criterion above, if translated from the terminology of the museum to that of the field, may thus be formulated as a certain degree of biological isolation from related groups.

After discussing freely hybridizing groups, sympatric ecological forms, plants, polyploidy[2] and asexuality, Huxley says, "Thus we must not ex-pect too much of the term species. In the first place, we must not expect a hard-and-fast definition, for since most evolution is a gradual process, borderline cases must occur. And in the second place, we must not ex-pect a single or a simple basis for definition, since species arise in many different ways" [p. 167].

The new geneticists tended, then, toward an eliminativist view on species, in what they perceived was the tradition of Darwin. Sure, the lineages split and this was real, but the rank of the splitting was manifold and had no universally common criteria that could be recognized. There was a division in the way the synthesis Darwinians approached species, which can be traced back to Darwin's own published ambiguity on the subject. Into this ambiguity of opinion came Ernst Mayr.

ERNST MAYR AND THE BIOSPECIES CONCEPT

Mayr was a German ornithologist who had left Germany well before the World War II and came to America [Hull 1988c: 66f.] to the American Museum of Natural History and then to Harvard, after spending a number of years in New Guinea and the Solomon Islands studying bird populations and distributions. He was motivated to address the "species problem" because of the publication of another book, opposed to Dobzhansky's approach, by geneticist Richard Goldschmidt [1940], who became Mayr's *bête noire* for many years to come. Goldschmidt proposed that species evolved in a single step, through macromutations involving chromosomal repatterning to form "hopeful monsters" [pp. 390–393] (it should be clear now that the term *monster* here refers to a sport or sudden variation from the type) in what is often called "saltation" (in other words, the opposite of *natura non facit saltum* quoted by Darwin). Goldschmidt repeatedly referred to species being separated by "bridgeless gaps" [cf. p. 143], a phrase he took from Turesson [1922b: 100], ignoring the fact that Turesson then went on to give a Darwinian account of species formation. Goldschmidt rejected the Darwinian idea that subspecific races were incipient species entirely, which seems to have motivated Mayr's ire and to have informed his allopatric account of speciation later.

Mayr was invited to give a series of talks on speciation as part of the Jesup Lectures in 1941 at Columbia University's Zoology Department. He was later invited to publish sufficient material to fill an entire volume for Columbia University Press after the other lecturer, Edgar Anderson, fell ill [p. xvii, of the new introduction to the 1999 reissue of his 1942 work]. The result was the single most widely referred-to volume of the synthesis.

Basically this work is a discussion at length of the modes of speciation according to the best knowledge of the day, and much of what Mayr discussed remains valid. In the case of "ring-species," for example, his discussions remain the canonical ones [pp. 180–185].[3] He also introduced several terms, including *allopatry* [p. 149] and *sibling species* [p. 151], which have worn well. But for our purposes, the most important aspect of the book is that here Mayr popularized the definition of species he had already given in his [Mayr 1940]:

> A species consists of a group of population which replace each other geographically or ecologically and of which the neighboring ones intergrade or interbreed wherever they are in contact or which are potentially capable of doing so (with one or more of the populations) in those cases where contact is prevented by geographical or ecological barriers.
>
> Or shorter: Species are groups of actually or potentially interbreeding natural populations, which are reproductively isolated from other such groups. [p. 120]

Mayr contrasted this with the "typological" conceptions of taxonomy at the time, especially the "morphological species concept," as he called it (for the first time anyone had, so far as I can tell), as well as the "practical species concept" (a version of Regan's definition), the "genetic species concept" (homozygous populations, which he attributes to Lotsy), and one based on sterility. He called it the "biological species concept"; and so the proliferation of general "species definition" names began. Mayr included Dobzhansky's definition under this rubric, but says of it that "[t]his is an excellent description of the process of speciation, but not a species definition. A species is not a stage of a process, but the result of a process" [p. 119].

Hence, he proposed the definition given here as a practical compromise, allowing the systematist the judgment call, since (as is noted later by critics of the concept) one cannot use reproductive isolation as a test in many cases, not least in allochronic and allopatric populations. Furthermore, "[t]he application of a biological species definition is possible only in well-studied taxonomic groups, since it is based on a rather exact knowledge of geographical distribution and on the certainty of the absence of interbreeding with other similar species" [p. 121]. Moreover, while it works for "bisexual organisms" (sexual species), it fails, he notes, for "aberrant cases" like protozoans and plants that are either unisexual (asexual) or freely hybridizing [p. 122]. The remainder of the book discusses speciation processes, and Mayr presents a Wagnerian view that geographic isolation is a precondition for the formation of new species [pp. 154–185]. Sympatric species are reproductively isolated absolutely ("*otherwise they would not be good species*") [p. 149, italics in original], but the gaps that separate allopatric species are "*often gradual and relative, as they should be, on the basis of the principle of geographic speciation*" [p. 149]. He considers sympatric speciation as a possibility but concludes:

> Darwin thought of individuals when he talked of competition, struggle for existence among variants, and survival of the fittest in a particular environment.[4] Such a struggle among individuals leads to a gradual change of populations, but not to the origin of new groups. It is now being realized that species originate in general through the evolution of entire populations. If one believes in speciation through individuals, one is by necessity an adherent of sympatric speciation, the two concepts being very closely connected. However, fewer and fewer situations are interpreted as evidence for sympatric speciation, as it is realized more and more clearly that reproductive isolation is required to make the gap between two incipient species permanent and that such reproductive isolation can develop only under exceptional circumstances between individuals of a single interbreeding population. [p. 190]

This has a whiff of circularity. Mayr defines species as reproductively isolated populations formed in allopatry and absolutely distinct in sympatry. He then claims that sympatric variations cannot form reproductively isolated species because species are formed in allopatry since sympatric "species" have to be absolutely isolated. Of course, there is a lot more to it than this, and Mayr brings in all sorts of impressive empirical evidence, so the charge of plain circularity, once made, must be dismissed. However, this means that the strength of the biological species concept rests *entirely* on the absence of plausible *empirical* reasons to believe that reproductive isolation does not occur in sympatric populations. All it would take to undercut this definition of species, of course—or at any rate, Mayr's argument in its favor—is to find an unambiguous case of sympatric speciation. There is a case—cichlid fishes in various lakes in Africa, for instance, where Mayr wonders if these "species flocks" are evidence for "explosive" sympatric speciation [p. 215]. He rejects the idea on the grounds that the closest relatives of each species are not sympatric.[5]

Finally, Mayr considers the factors that cause species, since species are the "effect" of a process. He divides them into internal factors, such as genetic mechanisms, mutation rates, and the like, and external factors. Then he lists a number of subcategories, isolating mechanisms: geographic barriers that restrict random dispersal, ecological barriers, ethological factors, mechanical factors, and "genetic" and physiological factors [p. 237f.].

A case for which Mayr does not insist on allopatry is what he calls "instantaneous sympatric speciation" [pp. 190ff.], such as via polyploidy (the duplication of chromosomes and possible subsequent reduction to a diploid form of differing composition to the parental species), self-fertilization, parthenogenesis and so on, but he says this is "apparently rare, even where it is hypothetically possible" [p. 192].[6]

In later publications, Mayr introduced the notion of species being "non-dimensional"[7]:

> Noninterbreeding between populations is manifested by a gap. It is this gap between populations that coexist (are sympatric) at a single location at a given time that delimits the species recognized by a naturalist. Whether one studies birds, mammals, butterflies, or snails near one's home town, one finds species clearly delimited and sharply separated from all other species. This demarcation is sometimes referred to as the species delimitation *in a non-dimensional system* (a system without the dimensions of space and time). [Mayr 1970: 14f., italics in original]

He also adds the distinction between the "species category" and the "species taxon"—the former "designates a given rank in a hierarchic classification" [p. 13], while the taxon is "the concrete object of classification. Any such group of populations is called a *taxon* if it is considered sufficiently distinct to be worthy of being formally assigned to a definite category in the hierarchical classification. *A taxon is a taxonomic group of any rank that is sufficiently distinct to be worthy of being assigned to a definite category*" [p. 14, italics in original].

By the 1963 version, the biological species as defined by Mayr has become a reproductive community, an ecological unit, and a genetic unit:

> [S]pecies are reproductive communities. The individuals of a species of animals recognize each other as potential mates and seek each other for the purpose of reproduction. A multitude of devices insure intraspecific reproduction in all organisms. . . . The species is also an ecological unit that, regardless of the individuals composing it, interacts as a unit with other species with which it shares the environment. The species, finally, is a genetic unit consisting of a large, intercommunicating gene pool, whereas the individual is merely a temporary vessel holding a small portion of the contents of the gene pool for a short time. [Mayr 1963: 21]

And by 1970, the definition has changed to read, "Species are groups of interbreeding natural populations that are reproductively isolated from other such groups" [quoted in Mayr 1970: 12, italics in original]. Gone is the phrase "actually or potentially" from the 1942 edition; Mayr now thinks that it is only in sympatry ("with respect to sympatric and synchronous populations," p. 13) that we can tell for sure that two organisms are distinct species. The definition is "biological," he says,

> not because it deals with biological taxa, but because the definition is biological. It utilizes criteria that are meaningless as far as the inanimate world is concerned.
>
> When difficulties are encountered, it is important to focus on the basic biological meaning of the species: *A species is a protected gene pool. It is a Mendelian population that has its own devices (called isolating mechanisms) to protect it from harmful gene flow from other gene pools.* Genes of the same gene pool form harmonious combinations because they become coadapted by natural selection. Mixing the genes of two different species leads to a high frequency of disharmonious gene combinations; mechanisms that prevent this are therefore favored by selection. [Mayr 1970: 13, emphasis added]

The text emphasized here is surprising. From being the "effect" of a process in 1942, Mayr now treats species as a *mechanism* of protecting gene pools. Moreover, selection now plays a role in "protecting" species; previously species were not formed through selection, and reproductive isolation was a side effect of geographic isolation. Still, isolating mechanisms are still "potentially or actually" active in sympatry and have to be intrinsic mechanisms of the organisms, and not, for example, "geographic or any other purely extrinsic isolation" [p. 56].

Mayr's view of species seems not to have changed much since the 1970 volume.[8] He repeats it in several places [e.g., Mayr 1976, 1985, 1988, 1992, 1996]. To summarize his final position, let us consider the latest paper [Mayr 1996], where he claims that species are concrete describable objects in nature, that they are reproductively isolated even when there is "leakage of genes," and that the biological species concept (now abbreviated as BSC) is based on the properties of populations. Although Mayr always stressed the populational nature of species as a result of his insistence on genetic and morphological polytypy in species, over time there is an increasing emphasis on "populational thinking" in opposition to "essentialism" in his works. For instance, in the introduction to his history of biology [Mayr 1982: 45–47], he discussed this, citing Hull [1976] and Ghiselin [1974b], dividing Western thinking into two phases. The first phase was essentialism deriving from Plato, and the second, population thinking beginning with Leibniz's theory of monads but really taking root with the British animal breeders and Darwin and his contemporary systematists in Britain. One can take issue with the claim for Leibniz (as a Great Chain thinker, the sort of variation he admitted was scalar rather than distributional) and wonder why de Quetelet has been overlooked as the source of populational thinking and instead been called an essentialist [see, for instance, Krüger et al. 1990]. Most oddly, what is missing here and elsewhere, is Karl Popper. Popper had attacked what he called "methodological essentialism" as a malign heritage from Plato [Popper 1957a, 1957b, 1960], in particular in his *Poverty of Historicism*, section 10, where he set up *nominalism*—the doctrine that universal terms are mere labels attached to sets of things—opposed to "realism" or "idealism," which he renames as *essentialism*. Popper ascribes to Aristotle the problem this introduces into science:

> The school of thinkers whom I propose to call *methodological essentialists* was founded by Aristotle, who taught that scientific research must penetrate to the essence of things in order to explain them. Methodological essentialists are inclined to formulate scientific questions in such terms as 'what is matter?' or 'what is force?' or 'what is justice?' and they believe that a penetrating answer to such questions, revealing the real or essential meaning of these terms and thereby the real or true nature of the essences denoted by them, is at least a necessary prerequisite of scientific research, if not its main task. *Methodological nominalists*, as opposed to this, would put their problems in such terms as 'how does this piece of matter behave?' or 'how does it move in the presence of other bodies?' For methodological nominalists hold that the task of science is only to describe how things behave, and suggest that this is to be done by freely introducing new terms wherever necessary, or by re-defining old terms wherever necessary while cheerfully neglecting their original meaning. For they regard *words* merely as *useful instruments of description*. [Popper 1960: 28f.]

Popper's sympathies are clearly with the nominalists. Through Hull's seminal essay in 1965 on essentialism in taxonomy, Popper's distinction came to be widely accepted amongst taxonomists, and Mayr may be influenced either directly or indirectly by Hull.[9] It is unclear whether he was directly influenced by Popper's *Poverty*, but that work was a cause célèbre in its day, and since Mayr and Popper were both leading German-speaking academics in the English-speaking world, it would be surprising if someone as erudite as Mayr had not at least heard of Popper and his ideas.[10] In the 1982 history, Mayr cites Popper only for issues of theory falsification. Why is this? I conjecture that it may be due to the prominence given to Popper by cladists such as Farris, Wiley, Patterson, Platnick, and Nelson as a justification for cladistic senses of naturalness [cf. Hull 1988c: 129, 171, 197, 237–239, 247, 251–253, 268, and references cited there].[11] In any event, this is beyond our scope here. The point is that Mayr is assuming a nominalistic approach to species; the term (as a category) is a useful way of organizing real, concrete objects (the taxa).

He particularly criticizes another philosopher, Philip Kitcher [1989], for instance, for failing to appreciate the difference between biological populations and classes of nonliving (inanimate) objects [Mayr 1996: 266f.].

Of particular note is Mayr's version of the history of species concepts. He gives it again in this paper, but he has given various forms of it in his other writings [Mayr 1957, 1970, 1982, 1991]. According to this version [Mayr 1996: 266f.],

> [t]he biological species concept developed in the second half of the 19th century. Up to that time, from Plato and Aristotle until Linnaeus and early 19th century authors, one simply recognized "species," eide (Plato), or kinds (Mill). Since neither the taxonomists nor the philosophers made a strict distinction between inanimate things and biological species, the species definitions they gave were rather variable and not very specific. The word 'species' conveyed *the idea of a class of objects, members of which shared certain defining properties.* Its definition distinguished a species from all others. Such a class is constant, it does not change in time, all deviations from the definition of the class are merely "accidents," that is, imperfect manifestations of the essence (eidos). Mill in 1843 introduced the word 'kind' for species (and John Venn introduced 'natural kind' in 1866) and philosophers have since used the term natural kind occasionally for species. [italics inoriginal]

In many details this account is incorrect, as the preceding chapters show. A distinction *was* made in practice between living and nonliving things by many authors before the nineteenth century, Locke introduced the term *kind* for species, and typological accounts permitted variation in the "essential" characters of type to quite a degree before members of the type became monsters. Mayr seems insistent on finding "forerunners" to his own preferred conception. On the next page, he writes of "the morphological, or typological species concept":

> Even though this was virtually the universal concept of species, there were a number of prophetic spirits who, in their writings, foreshadowed a different species concept, later designated *[by Mayr, as it happens—JSW]* as the *biological species concept* (BSC). The first among these was perhaps Buffon (Sloan 1987), but a careful search through the natural history literature would probably yield quite a few similar statements.

This tendency to seek precursors for a favorite personal view is known among historians as the "Whig interpretation of history" [Butterfield 1931] or as "presentism" or "creeping precursoritis";[12] it is the importation of modern views into the past. At least since Collingwood [1946], such approaches to history as the progressive buildup to the modern day or some ideal state have been widely viewed askance by historians: "Bach was not trying to write like Beethoven and failing; Athens was not a relatively unsuccessful attempt to produce Rome; Plato was himself, not a half-developed Aristotle" [p. 329]—and, we might add, the writers on species in the period in question were not precursors to Ernst Mayr.[13]

There has *always* been a "biological" (reproductive) component to discussions of species as applied to the living world, which I have called the "generative" notion of species. Moreover, few of the writers adduced by Mayr as forerunners actually are presenting anything much like his view, as most of them include a clear morphological component in their conceptions. Nevertheless, Mayr's claim of E. B. Poulton and K. Jordan as "precursors" would seem to be fair, at least in terms of a similarity of views; given the number of times he cites them, he may even have them as direct antecedents—that is, they might have directly influenced *him*. Stresemann, as his teacher, clearly did [Winsor 2004; Chung 2003].

In his "point paper" on the biological species concept [Mayr 2000a] in the Wheeler and Meier volume [2000], Mayr repeats most of the previous paper. One thing he does add here is that "[t]he word *interbreeding* indicates a propensity; a spatially or chronologically isolated population, of course, is not interbreeding with other populations but may have the propensity to do so when the extrinsic isolation [is] terminated" [p. 17].

The shift in Mayr's thinking from "actually or potentially interbreeding" in 1942, to "actually or potentially operating isolating mechanisms" in 1963 to "propensity to interbreed" in 2000 is interesting. As a test of species status, potential anything is obviously useless unless it can be made actual (which is the basis for his insistence that only in sympatry are species fully determinable[14]). But clearly one wants to be able to say that I and the inhabitants of fifteenth-century England are the same species, even if I cannot interbreed with them *actually*, and so Mayr *must* introduce something like a propensity interpretation. However, "propensities to behave" are themselves no easier on the metaphysical eye than potentialities. They remain, in the end, conditional statements: a lump of sugar is soluble, if, when immersed in water, it dissolves [cf. Sober 1984: 76–78]. But if there were no possibility of immersing it in water, would it remain a soluble substance? We want to say so, but if we examine our intuitions, this is because we know of other lumps of sugar that they *have* dissolved, and by analogy with this lump, which has the same composition, and which we have no reason to think otherwise, so it, too, will dissolve. Propensities may also be interpreted as the average behavior of a reference class in certain conditions [Hájek 2003] but in this case it makes no sense to talk about the propensity of an individual case (that is, of me and a fifteenth-century woman in England). Either way, Mayr has a problem, with potentials or propensities.

Mayr's conception of species is, in the end, one of dispositions to behave in various ways. Henry IV and I are of the same species because we are of the same "substance," and were we in the same population, our genes could freely spread through it. But this is what Mayr wants to object to—this smacks of essentialism, though, as I have argued, it isn't. In the meantime, let us note that Mayr tries to avoid counterfactual claims of "*would* have interbred, if in the same (natural) population" with his insistence on sympatry for full species-hood. It is a problem he makes largely for himself, based on his conflation, I believe, of the epistemic and ontological aspects of being a species. In this, he is not alone.

MODERN DEBATES

We shall now review the broad species concepts presently in play. These are several classes of fundamental concepts, here divided into "Reproductive Isolation Concepts," "Evolutionary Concepts," "Phylogenetic Concepts," "Ecological Concepts," and a trash can category of "Other Concepts" [Cracraft 1997; Mayden 1997; Wheeler and Meier 2000; Hey 2001a; Mayden 2002]. From these, a number of subsidiary concepts are composed.

REPRODUCTIVE ISOLATION CONCEPTS

They opened Buffon again and went into ecstasies at the peculiar tastes of certain animals. . . .

They wanted to try some abnormal mating. . . .

They made fresh attempts with hens and a duck, a mastiff and a sow, in the hope that monsters would result, but quite failing to understand anything about the question of species.

This is the word that designates a group of individuals whose descendants reproduce, but animals classified as different species may reproduce, and others, included in the same species, have lost the ability to do so.

Gustave Flaubert [1976: 87][1]

The notion that species are kinds of organisms delineated not by the decisions of the classifiers but by the reproductive behaviors and results of the organisms themselves is an old one. As we have seen, it was suggested as part of John Ray's definition in 1688 and also by Buffon in 1748, but it goes back in the form of the generative conception of species to the time of the Greeks. To a greater or lesser degree, it has been a component of nearly all species concepts since Linnaeus (his sexual system implicitly required reproductive isolation), and even now, it is a key component [Cracraft 2000] of phylogenetic species concepts and most other operational definitions.

The generative conception of species has been applied to living beings effectively back to Epicurus and the neo-Platonists. That is, there has *always* been a requirement not only of constancy of form in definitions of living species but of the reproduction of form. This is surprising, since the implication or tacit assumption of many discussions, such as [1982] or Hull's [1965, 1988c], has been that species had been for much of the history of the concept arid definitional constructs based on "essences." There can be no doubt that essence *has* played a role in species concepts. However, as we have seen, at least since Locke there has always been a tacit distinction between the *real* essence of a species (that is, what causes a species to *be* a species, its "real constitution") and the *nominal* essence (that is, how we know the species, describe it, define it, and apply a name to it).[2] It is not clear that "essence" in these cases plays a definitional role—the Real Essence is not, by definition, definable.

Modern views of species are widely known and discussed, so we shall be brief in covering them, except for the ways that the most well-known reproductive isolation concept, the class of which we shall call for brevity "isolation concepts"—which is of course Ernst Mayr's biological species concept, or *biospecies*—developed. Also, there is considerable confusion and ambiguity in the phylogenetic concepts, so this topic will need discussion. I have attempted to be comprehensive, though, and provide all the current twenty-five or so species concepts on offer [building on Mayden 1997] since Dobzhansky introduced the issue into the synthesis debate.

Recognition Concepts

There have been numerous conceptions of species following Mayr that are fully, or partially, isolationist. The views of Hugh Paterson [1985, 1993] in particular have influenced Niles Eldredge [1989, 1993] and Elisabeth Vrba, among others [Lambert and Spencer 1995]. Paterson's version of the biospecies concept requires that organisms share a mating system, which he terms the "Specific-Mate Recognition System," or SMRS. He intends this to apply to plants, animals, and other organisms, so the term *recognition* should be taken in the same way the term *selection* is—without voluntaristic or cognitive implications. Paterson 1985: 25] defines a species thus: "We can, therefore, regard a species as that most inclusive population of individual biparental organisms which share a common fertilization system" [italics in original]. He refers to this as the "recognition concept." Mayrian and Dobzhanskyan concepts he calls "isolation concepts." Since Paterson's version applies, as did previous isolation concepts, only to fully sexual and gendered (anisogamous) organisms [p. 24], it follows that like them, he does not regard nonsexual organisms as forming species. Paterson's criticisms of the isolation concept include its being teleological, since species are seen as "adaptive devices" [p. 28], but this is hardly fair. Biospecies are not, on Mayr's account, adaptive any more than higher taxa, and on his account they are formed through isolation and subsequent local adaptation, not *by virtue of* their adaptations. The main difference between the isolation conception of Mayr and that of Paterson is well noted by another contributor to that volume, Scoble [1985], who notes that "the BSC can apply to only *actually*, not potentially, interbreeding groups of organisms. However, if the recognition concept of species . . . is accepted . . . then we may be able to directly compare at least some of the very characters involved in mate recognition in allopatric and allochronic populations" [p. 33].

However, this does not follow, since we cannot say whether the mate recognition sequences are compatible enough to form viable progeny, either occasionally or repeatedly enough to make them count as the "same" species, until we have actually managed to test it in the lab and the wild. Scoble also notes the problem of uniparental species and suggests that a homeostatic view might pertain. Since the SMRS is only one kind of homeostatic mechanism, there is no reason to restrict specieshood to sexual organisms only.

Genetic Concepts

We have several genetic concepts of species, ranging from Dobzhansky's comment that species are "the most inclusive Mendelian population" and Carson's comment that a species is a "field for gene recombination" [Carson 1957],[3] to fully worked-out genetic conceptions like Templeton's and Wu's.

Templeton's view is that a species is "The most inclusive group of organisms having the potential for genetic and/or demographic exchangeability" [1989: 25]. Templeton considers two criteria for being a species: genetic and demographic exchangeability. The first he specifies as "the factors that define the limits of spread of new genetic variants through *gene flow*," and the second as "the factors that define the fundamental niche and the limits of spread of new genetic variants through *genetic drift* and *natural selection*" (see his table). A sexual species requires both criteria [see the discussion in Wilkins 2007a]. He sees the cohesion concept as sharing a lot with the evolutionary species concept in this respect. These mechanisms generate a cohesive group, and the concept includes asexual taxa (purely in terms of demographic exchangeability) as well as sexual taxa (a mix of both). Nevertheless, Templeton does not indicate by what the level of species is indicated. His is more a matter of identifying what mechanisms generate species, whatever that level may be. Under both mechanisms, the spread of genetic variants through populations is the *sine qua non* of species rank.

As Ghiselin [1997: 113] observes, Templeton's conception resembles Mayr's in that species form as the result of genetic revolutions that constrict exchangeability, but he also favors Paterson's views on mate recognition. "Casting the definition in terms of one gene to be exchanged for another makes it roughly the same as the biological species definition." The requirement for demographic exchangeability means that organisms as units can replace each other in genetic populations, even if they do not do so in actuality.

James Mallet offered up "a species concept for the modern synthesis" [Mallet 1995; cf. also Mallet 2000, 2001; Dres and Mallet 2002; Naisbit et al. 2002; Beltran et al. 2002], the *genotypic cluster species concept* (GCSC), in which species were identified as "identifiable genotypic clusters" [Mallet 1995: 296]. Although a genotypic notion, in many ways it resembles the phenetic concept (discussed later) in which instead of morphological variables, the lack of intermediates lies in single genetic loci and ensembles of multiple loci. Like the biospecies concept, the GCSC applies only to populations in sympatry or parapatry [Brower 2002], and it lacks a nonarbitrary level of similarity or isolation of genetic alleles to specify species-hood. In many ways it appears to be a genetic version of the diagnostic species concept described later.

Wu's [2001] views are harder to pin down exactly; he may not, in fact, have a distinct species concept at all. He claims that reproductive isolation (RI) has involved the entire genome under the BSC but that it instead involves genetic isolation of *adaptive genes and gene complexes*, in particular what he calls *speciation genes*. Even if there remains gene exchange for the nonadaptive genes, if there is RI for the adaptive genes, good species have evolved. His view relies on speciation itself, the process, resulting in some degree of genome isolation. Species are defined in terms of their genes having reached a particular level of RI, either Stage III, where populations have diverged sufficiently that they will not fuse in sympatry, or Stage IV, where populations are entirely isolated and will not mix at all genetically. Speciation genes have become a topic of interest to researchers with the rise of sympatric speciation models and examples [Butlin and Ritchie 2001; Orr and Presgraves 2000], but Wu's genic conception is within the traditional genetic and biospecies concepts [Vogler 2001; Van Alphen and Seehausen 2001; Rieseberg and Burke 2001; Bridle and Ritchie 2001]. The mode of speciation as a way to identify the nature of species is the subject of one of my papers [Wilkins 2007b].

Jerry Coyne and Allen Orr: The BSC Revivified

In a major review of research into speciation, Jerry Coyne and Allen Orr present a revised version of the BSC in which they make a number of concessions to criticisms and in which they defend against some [2004, chapter 1]. Coyne and Orr's version, which I will here call the *limited BSC*, allows for limited introgression, does not insist on integration of gene complexes, permits other definitions for asexuals and paraphyly of species, and treats ecological differentiation as necessary to the persistence of species (in sympatry, at any rate) but not to the definition based on reproductive isolation. It allows that nongenetic isolation can form species (for example, intracellular infection by *Wollbachia*) and, further, that good species might later hybridize to form a new species. They accept Mayr's definition: "species are groups of interbreeding natural populations that are reproductively isolated from other such groups" [1995: 5], although given their qualifying of that concept, it should perhaps read *mostly* interbreeding and isolated.

EVOLUTIONARY SPECIES CONCEPTS

Evolutionary species concepts derive from attempts to deal with the time dimension, something that the biological species concepts of Mayr and Dobzhansky tend to avoid—for them, species exist at a given time horizon, and over evolutionary time, of course, they can change and split. Palaeontologist George Gaylord Simpson proposed a definition: "An evolutionary species is a lineage (an ancestral-descendant sequence of populations) evolving separately from others and with its own unitary evolutionary role and tendencies" [1961: 153]. Earlier, in a classic paper [Simpson 1951], he had expressed it slightly differently: "a phyletic lineage (ancestral-descendant sequence of interbreeding populations) evolving independently of others, with its own separate and unitary evolutionary role and tendencies, is a basic unit in evolution" [quoted in Ghiselin 1997: 112f.][4]

According to Cain [1954: 111], Simpson characterized the intergrading forms of an evolutionary species as "transients." Species are thus a number of things: they are *populations* that form *phyletic lineages* through *interbreeding* and that have *independent evolutionary roles* and *independent evolutionary tendencies*. These properties are reiterated in a later version of the evospecies concept of E. O. Wiley: "A species is a single lineage of ancestral descendant populations of organisms which maintains its identity from other such lineages and which has its own evolutionary tendencies and historical fate" [1978: 18]. By 2000, this has ceased to be a "definition" and is now a "characterization": "An evolutionary species is an entity composed of organisms that maintains its identity from other such entities through time and over space and that has its own independent evolutionary fate and historical tendencies" [Wiley and Mayden 2000: 73]. The novel elements here include the "entification"[5] of evospecies in place of the population stipulation, to accommodate asexual organisms [p. 75], and the inclusion of time and space to indicate the processual nature of the concept. However, one thing that all the evolutionary concepts fail to do adequately is to specify what counts as "independence." If a parasitical species coevolves with its host, which is commonly the case, are they still independent? And, of course, there is a metaphorical problem with "fate," given the universal view that evolution is not predetermined, but we can assume that Wiley and his colleagues understand that; this merely specifies that the outcomes of evolution for any one species are unique to that species. In a sense, this is a matter of evolving a unique set of traits.

Evolutionary species concepts have been assumed by some critics to imply a gradual and constant rate of evolution [Ayala 1982; Eldredge and Gould 1972; Gould 1982]. This need not be the case, however, as Simpson's own classical work on evolutionary rates indicates [Simpson 1944; Gould 1994], but it is also sometimes held that species change over the entire course of their duration. Opposing this, the punctuated equilibrium theorists [Eldredge and Gould 1972; Eldredge 1985; Eldredge et al. 1997; Gould and Eldredge 1977; Gould 2002] have argued that species tend to remain stable once evolved. If correct—or rather, *when* correct, for it is now accepted to be the case for many species if not all—the "fate" of the species involves the stasis of the unique set of traits once achieved.

Evolutionary species concepts, as with some phylogenetic species concepts, tend to adopt the metaphysical species-as-individuals thesis of Ghiselin and Hull (Ghiselin 1974a, 1988, 1997; Hull 1976, 1978, 1981, 1992b). As Wiley and Mayden note, "Evolutionary species are logical individuals with origins, existence, and ends" [p. 74]. However, if species are correctly thought of as *logical* individuals, they still need not be *historical* individuals. There are three different notions under the term *individual* in the species concept literature, and they ought to be kept distinct: individuals as *metaphysical particulars* (i.e., not universals or natural kinds), individuals as *coherent functional objects* like organisms, and individuals as clusterings of properties in a phenomenally salient manner. Despite comments like those of Ghiselin in his 1997 book *Metaphysics and the Origin of Species*, where he takes metaphysical particularity to imply functional coherence, these are not the same thing. The volume bounded by a sphere of circumference 1,000 kilometers from a point 450 kilometers directly above Hawaii is not a coherent functional system, but it is a particular. Nor does either sense imply that the functional object or particular is phenomenally salient.

Evospecies, as we might call these entities, are something of a hybrid notion, to my mind. They are phyletic objects, but they are often presented as achieving grades of organization, and often run with evolutionary systematics (itself a hybrid, conjoining phylogenetic and adaptive conceptions of classification).

Lineages

A version of the evospecies is Kevin de Queiroz's general lineage conception. The key term *lineage* is taken from Hull's metaphysics of [Hull 1980, 1981, 1984c, 1988a, 1988c, 1989b], based on Simpson's evolutionary species definition [1951, 1961], revised by de Queiroz [1998, 1999] (see figure 8). It represents the *generative* aspect of species that we saw from early in the history of the concept. It broadly means

> a series of entities forming a single line of ancestry and descent. . . .
> Lineages in the sense described above are unbranched; that is, they follow a single path or line anytime an entity in the series has more than one descendant. . . . Consequently, lineages are not to be confused with clades, clans, and clones—though the terms are often used interchangeably in the literature. [de Queiroz and Donoghue 1990: 50]

> [A] lineage (an ancestral-descendant sequence of populations). [Simpson 1951]

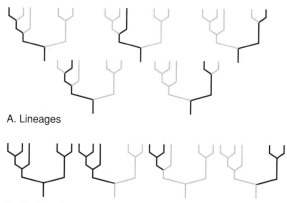

A. Lineages

B. Clades, clans, or clones

Figure 8 De Queiroz's conception of lineages, redrawn from de Queiroz [1999: 51]. The caption reads: "Lineages contrasted with clades, clans, and clones. . . . All of the branching diagrams represent the same phylogeny with different lineages highlighted in (A) and different clades, clans, or clones highlighted in (B). Notice that the lineages are unbranched and partially overlapping, whereas the clades, clans or clones are branched and either nested or mutually exclusive. Additional (partial) lineages can be recognized for paths beginning at various internal nodes."

A problem with this is that while every species is a lineage (that is, forms generative sequences), not every lineage forms a species. Essentially a lineage is any part of the evolutionary tree that is unbroken (see figure 8), but we know that species have within them subordinate lineages, of populations, of genes, and of kin groups. As we shall see with phylogenetic species concepts in the next chapter, this leads to problems in specifying what level of lineage one wants to call a species.

For example, harbor seals and albatrosses [Burg and Croxall 2004; Burg 1999] turn out to have extensive within-species variation that form lineages. The decision of the level of diversity of lineages is informed by other considerations, including what the taxonomist considers a "significant" amount of genetic divergence or not. But this means that a lineage, or indeed an evolutionary species, is determined in the end by phenetic criteria—the rank depends on arbitrary (but not necessarily subjective) lines of divergence in the characters or genetic matrix used.

One attempt to deal with this is the genealogical concordance species concept of Avise and Ball [1990]. De Querioz [2007] considers this a phylogenetic conception of species, but in my opinion it is better thought of as an evolutionary conception. It is basically the view that population subdivisions concordantly identified by multiple independent genetic traits constitute the population units worthy of recognition as phylogenetic taxa. Relying on genetic coalescence theory, in which genetic lineages are traced back to original populations [Avise and Wollenberg 1997; Hey and Wakeley 1997; Knowles and Carstens 2007], it considers a species to exist when it shares a number of genetic coalescents.

PHYLOGENETIC SPECIES CONCEPTS

Phylogenetic conceptions of species are very similar in some ways to the classifications of the older logic of division discussed earlier. As Nelson and Platnick [1981] note, Porphyry's tree is strongly reminiscent of the later apomorphy/plesiomorphy distinction made by Willi Hennig [1966]. Hennig distinguished between derived and underived states in taxonomy. *Plesiomorphic* characters are those in a monophyletic group (a clade) from which the transformations begin, and *apomorphies* are those derived from them in evolution [p. 89]. An apomorphy of a group can be a plesiomorphy of a clade contained within that group. It follows that in Hennig's classificatory scheme that what makes a plesiomorphy and an apomorphy is not absolute rank, but its relative position in a cladogram. In the older terminology we have already discussed, they are relative to the differentiae, the predicates and the properties those predicates denote that make the differences between taxa. Hennig, like Martianus, has no system of absolute taxonomic levels, although Hennig did also attempt to develop such an absolute set of ranks based on time of evolutionary divergence [Hennig 1966: 184–192]. There are just taxa, and they are arrayed in a flexible local hierarchy [Nelson 1989]. In the classical logic, the base-level taxonomic rank—the *infimae species*—was a taxic entity that was not itself the genus of any other species and which contained only individuals. Likewise, Hennig has terminal taxa, and these he calls *species*, following Linnaean tradition. Where the infimae species and the Hennigian species differ from the Linnaean species, however, is that the former are derived from the general group being sequentially divided into subgroups on the basis of characters shared (a single dichotomous key in the early naturalists' practice, on an implied parsimony criterion of many characters for Hennig), while Linnaeus assumed fixed taxon ranks. Linnaean species are an absolute rank, and so also are the higher taxa they comprise.

Species concepts based on the phylogeny of the groups of organisms are called "phylogenetic," but there are several *phylospecies* concepts, as I will call them, and there are several sub-versions of them in turn. All proponents of phylospecies concepts claim both Darwin and Hennig as their inspirations, but it is arguable how closely each of the modern views relate to those initial expressions of classification of taxa by descent.

In some ways, phylogenetic classification is an outgrowth of the ideas expressed by Haeckel and others during the late nineteenth century. Haeckel coined the term *monophyly*, which Hennig later appropriated and more strictly defined. This is a critical notion in the context of phylogenetic systematics or, as it is popularly known, cladistics. In Hennig's definition, a monophyletic group is an ancestral species and *all* of its descendant species.[6] Under more formally defined notions of phylogeny, sometimes called *pattern cladism*, a monophyletic group is a proper subset of some set of taxa, without necessarily implying a particular historical relationship. On this account, a monophyletic group is definable in terms of it having unique characters that are not shared by other taxa. These are called *apomorphies* in Hennig's terminology, or "derived characters." This is a relative term: an apomorphy for one set of taxa is a *plesiomorphy* (an underived, or primitive, character) for a more inclusive set of taxa. If one or more taxa share an apomorphy, it is called a *synapomorphy*; and if only a single taxon carries an apomorphy, it is called an *autapomorphy*.

There are fundamentally three phylospecies concepts (figure 9). The first, defined initially by Hennig, is sometimes called the *Hennigian species concept* [Meier and Willmann 2000]. It rests on a conventional decision Hennig made to be consistent with his broader philosophy of classification, which we shall call the *Hennigian convention*. On this account, if a new species arises by splitting from a parental species, then we shall say that the parental species, no longer being monophyletic (that is, now being paraphyletic), has become extinct, and there are now two novel species.

The second phylospecies concept we might call the *synapomorphic concept*. This further divides into the monophyletic concept of Mishler et al., and those who merely apply synapomorphic criteria. A species is under this account, one in which there is no further division or taxonomic subdivisions—it is the unit of phylogenetic analysis. Mishler's version adds the view that species must be monophyletic.

The third phylospecies concept, which I shall call the *autapomorphic concept*,[7] a phylogenetic version of what we might term Diagnostic Species, is derived from the work of Donn Rosen [Rosen 1979]. In various

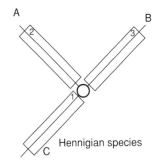

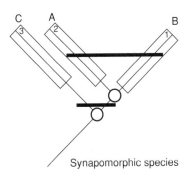

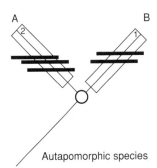

Figure 9 Three types of phylospecies concepts.

Hennigian species (top) are formed at speciation and are extinguished at the next speciation event or extinction. The cladogram is a history as well as a classification.

Synapomorphic species (middle) are terminal nodes in a cladogram or evolutionary tree specified by synapomorphies. They are recognized by a lack of further discrete patterns of ancestry and descent (i.e., by having no further synapomorphies).

Autapomorphic (diagnostic) species (bottom) are the smallest terminal nodes in a cladogram, marked by their autapomorphies, or unique set of characters depending on the authors. All three conceptions may also specify that species are either monophyletic or not.

versions, it tends to rely on the diagnosability of taxa [Cronquist 1978], or, as we may otherwise say, it is a largely epistemological notion of species. All the autapomorphic concepts rely on a species being the "terminal taxon" in a cladogram. Some proponents apply an evolutionary exegesis to this, while others restrict it to a diagnostic relationship, a distinction often referred to as the ontology—epistemology aspects of species [cf. Wheeler and Meier 2000]. Of course, a phylospecies of one kind can also be a phylospecies of another. The division of the phylospecies into two main kinds, ignoring Hennigian species for the moment, is a reflection of the larger taxonomic debates. These raise the following questions: (1) should taxonomic classification proceed in terms of descent alone or on the basis of similarity (cladism versus gradism)? and (2) if classification rests on clades, are homologies (apomorphies) indicators of history, or are they patterns that are evidence in favor of a historical reconstruction but not themselves a model of evolution? Briefly, this is the distinction between ontological and epistemological notions of classification again.

Those who take the epistemological classification position (grouping in terms of synapomorphic relationships only) often fall under the rubric of pattern cladists, although this is by no means necessarily so. Mishler and Theriot [2000], for example, take synapomorphies as indicating relations at a time, synchronically, to avoid time paradoxes (in which a species becomes paraphyletic after the speciation of daughter species, if it is treated as a sister taxon). Only tokogenetic relations are treated as diachronic relations under this view.

Those who take the diagnosis of monophyly of a group to give a direct hypothesis of evolutionary history are the so-called orthodox, or "traditional" cladists, called "process cladists."[8] Process cladism tends to treat taxa as relationships between organisms, while pattern cladism tends to see taxa as composite entities of which organisms are members. In this regard, the process orthodoxy is more closely allied to some aspects of the evolutionary species concepts of recent times.[9]

Phylogenetic approaches to species are founded on one of three criteria: synapomorphy, autapomorphy, or prior phenetic identity as OTUs. Roughly, synapomorphy acts as the classical notion of genera, autapomorphy acts as the classical notion of differentia; and in the case of phenetic OTUs as inputs into a cladistic analysis, species here are types that are known prior to the formal division. Therefore, it is something of a misconstrual to think that there are only two "phylospecies" concepts besides the Hennigian account.[10]

Hennigian, or Internodal, Species

In the work that began the cladistic revolution and that remains the source for much cladistic thinking, so clearly and consistently was it expressed,

Hennig treated species in an unusual manner [Hennig 1966: 28–32]. He begins by defining "semaphoronts," or "character bearers," as the elements of systematics. These are effectively stages or moments in the life cycle of organisms of the taxon. Individuals themselves—or, more exactly, the ontogenetic life cycles of individuals—are related through reproductive relationships he called "tokogenetic" relationships. When tokogenetic relationships begin to diverge, they form species, which arise

> when gaps develop in the fabric of the tokogenetic relationships. The genetic relationships that interconnect species we call phylogenetic relationships. The structural picture of the phylogenetic relationships differs as much from that of the individual tokogenetic relationships as the latter does from the structural picture of the ontogenetic relationships. In spite of these differences in their structural pictures, the phylogenetic, tokogenetic, and ontogenetic relationships are only portions of a continuous fabric of relationships that interconnect all semaphoronts and groups of semaphoronts. With Zimmermann we will call the totality of these the "hologenetic relationships." [p. 30]

Hennig illustrated this with a now-famous diagram (figure 10).

Hennig is sometimes read as asserting that species cease to exist when they are divided—that is, when the hologenetic relationships cease to be a single set (the species sets are represented in the diagram by the large ellipses). Although Hennig assumes that species are reproductive communities of harmoniously cooperating genes, as Dobzhansky had said, and that species are reproductive groups, as Mayr had said, Hennig nevertheless assumes that species are phylogenetic lineages. In fact, he says that species are defined over spatial dimensions as well, as "a complex of spatially distributed reproductive communities, or if we call this relationship in space 'vicariance,' as a complex of vicarying communities of reproduction" [p. 47]. The sort of vicariance he has in mind includes trophic replaceability [pp. 49–50] but also other ecological dimensions, including temperature races.

But the most significant aspect of Hennig's definition of species lies in the temporal dimension [pp. 58–60], where he notes that species are to be delimited by events of speciation: "The limits of the species in the longitudinal section through time would consequently be determined by two processes of speciation: the one through which it arose as an independent reproductive community, and the other through which the descendants of this initial population ceased to exist as a homogenous reproductive community" [p. 58]. Transformations of the

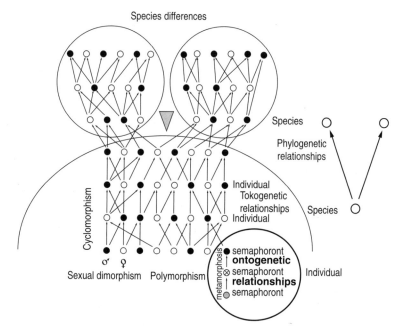

Figure 10 Hennig's view of systematic relationships. The larger circles are the "hologenetic" relationships that represent species. Redrawn from Hennig [1966: 31].

species morphology and genetic composition within these two events do not affect the identity of the species, because the species is defined here as a homogenous reproductive community. In effect, Hennig is taking the biospecies isolationist concept to its limits. Species are extinguished at the next speciation event. In his earlier work in German [Hennig 1950: 102], he was even more concise: "When some of the tokogenetic relationships among the individuals of one species cease to exist, it disintegrates into two species and ceases to exist. It is the common stem species of the two daughter species" [translated in Meier and Willmann 2000: 30].

This has become known as the *Hennig Convention*. It has been strongly criticized from all sides, by biospecies proponents, evospecies proponents, and other phyogeneticists [see the citations in Meier and Willmann 2000: 31], not least because it seems to be an arbitrary way to delimit species taxa. Mayr, for example, takes Hennig to be making a substantive claim on the ways species are formed at speciation, and he criticizes it on the basis that the "parent" species can remain unchanged even though it is no longer monophyletic. But it seems to me that the critics have overlooked the most charitable interpretation of the Hennig Convention—it is a convention about *naming* and *denotation*. In short, the *name* of a species is extinguished at speciation. This follows from Hennig's views about the task of systematics. Using (and citing) Woodger and Gregg [see the excellent discussion in Wheeler and Platnick 2000] and the views of Woodger [1937] in particular about sets in classification, Hennig strives to ensure that there is no ambiguity of reference in the sets named in systematics. Since as soon as a set is divided there is ambiguity, which of the two resultant sets is being referred to by a prior name, Hennig proposes to extinguish the now-ambiguous name and create two new ones. However, he seems to equivocate over whether or not they are new *entities*.

The Hennigian concept of species has been recently expanded and defended by several people. Meier and Willmann [Meier and Willmann 2000: 31; cf. Willmann 1985a, 1985b, 1997] have proposed a modified Hennigian species concept: "Species are reproductively isolated natural populations or groups of natural populations. They originate via the dissolution of the stem species in a speciation event and cease to exist either through extinction or speciation" [quoted in Willmann 1985a: 80, 176]. Like Hennig, and most proponents of the biospecies and other isolationist conceptions, they reject the use of a single taxonomic category such as *species* to apply to asexual ("uniparental") organisms. Instead, they call them "agamotaxa."

A set-theoretic model of Hennig's concept has been provided by Kornet [1993a, 1993b] under the title of "internodal species concept," who formalizes the notion of internodality (INT) in a cladograms in terms of the tokogenetic relations being permanently divided, so that any individual in the hologenetic group has a "gross dynastic relationship" (GDYN) with any other individual in that group. Type specimens fall out as an appropriate individual to start the GDYN analysis to establish the INT relation. Kornet thus makes the relation of type specimens to the rest of the species scrutable in set theoretic terms.

The Hennig account is fundamentally a biospecies concept. Hennig himself accepted that species were reproductively isolated, and the criteria used for identifying the relevant phylogenetic edges of the cladogram are simply those of the biospecies. We could therefore say that it is better considered to be considered under that rubric and that the issue of "extinction" of species at cladogenesis is one of the reference of taxonomic names. In short, the "extinction" is a taxonomic extinction.

Synapomorphic Species

There are many other conceptions that their authors refer to, or which are referred to by others, as "phylogenetic" conceptions of species. It is not clear that these form a natural class of conceptions. I believe there are two basic approaches that derive from the cladistic terminology, but it is not therefore the case that all authors in one or the other classes agree or that cladistic terminology resolves the differences between them.

Synapomorphic species are largely due to the advocacy of Brent Mishler, although earlier Joel Cracraft had defined the species taxon as "the smallest diagnosable cluster of individual organisms within which there is a parental pattern of ancestry and descent" [1983: 170]. Mishler and Brandon [1987] summarize their monophyletic version [Mishler and Donoghue 1982]:

> A species is the least inclusive taxon recognized in a classification, into which organisms are grouped because of evidence of monophyly (usually, but not restricted to, the presence of synapomorphies), that is ranked as a species because it is the smallest "important" lineage deemed worthy of formal recognition, where "important" refers to the action of those processes that are dominant in producing and maintaining lineages in a particular case. [Mishler and Brandon 1987; p. 310 in Hull and Ruse 1998]

They redefine *monophyly* in such a way as to be able to include species: "A monophyletic taxon is a group that contains all and only descendants of a common ancestor, originating in a single event" [p. 313 in Hull and Ruse 1998].

De Queiroz and Donoghue, on the other hand, treat species as systems that may not be monophyletic, and indeed may be paraphyletic if a species has split from it, in a parallel with cohesive and functional individuals who lose cells and reproduce [de Queiroz and Donoghue 1988, 1990]. They therefore exclude asexuals and indistinct populations that are not assignable from being members of species. Mishler and Theriot [2000], extending the monophyletic conception, include asexuals [p. 52f.] largely on the grounds that asexual taxa do not markedly differ in overall phylogenetic nature from sexuals—the number of autapomorphies, for example, are similar in both cases (one does wonder, though, if this is due more to the application of phylogenetic classification than anything else).

Synapomorphic species are usually based on the historical, actual lines of ancestry and descent that are represented in a cladogram. As a phylogenetic taxon, a species is grouped by the synapomorphies shared by organisms that indicates monophyly [Mishler and Theriot 2000: 47]. It is a phylogenetic or cladistic replacement for the evolutionary species concept of Simpson. Mishler and Theriot [2000: 46–47] give the following broad definition:

> A species is the least inclusive taxon recognized in a formal phylogenetic classification. As with all hierarchical levels of taxa in such a classification, organisms are grouped into species because of evidence of monophyly. Taxa are ranked as species because they are the smallest monophyletic groups deemed worthy of formal recognition, because of the amount of support for their monophyly and/or because of their importance in biological processes operating on the lineage in question.

Here, the monophyletic conception is explicit. The dual nature of the epistemic and the ontological aspects of species are clearly expressed, and the rank of *species* is restricted to "biologically important" lineages. Their version of the synapomorphic concept allows for reticulation as a general problem in classification not merely restricted to species taxa.

De Queiroz and Donoghue [1988, 1990], on the other hand, do not think that species have to be monophyletic, because monophyly of populations does not offer a way to specify what the base rank is; and because species evolve from ancestral populations, this will leave the species from which the ancestral population derived as paraphyletic. Of course, the monophyly spoken of here is somewhat different from the monophyly of the Mishler et al. version; this one is based on populations as the base entities; the Mishler et al. version is based on phylogenetic lineages. And both conceptions converge on a similar solution—species are regarded as singular phylogenetic *lineages*, which is de Queiroz's later conception of a cohesive object or group over phylogeny [de Queiroz 1998, 1999]. The actual answer of the earlier paper, however, is that there is no single definition of species that will "answer to the needs of all biologists and will be applicable to all organisms" [quoting Kitcher 1984: 309], although they reject the sort of pluralism I have proposed [see Wilkins 2003].

Autapomorphic (or Diagnostic) Species

Diagnosis of species has always been involved in the debate, but few, if any, have until recently suggested that diagnosis is sufficient, apart from (and probably not even there) the so-called taxonomic species concept. Cronquist [1978: 3] provided one of the first such conceptions: "Species are the smallest groups that are consistently and persistently distinct, and distinguishable by ordinary means." Some [e.g., Ghiselin 1997: 106f.] accuse Cronquist of presenting a "subjective" concept, but it all hinges on what "ordinary means" means. At one time, the use of a microscope was reviled (e.g., by Linnaeus); now assays for specificity ranging from molecular data to morphometric and acoustic traits are considered ordinary practice.

More common phylogenetic diagnostic concepts, though, arise from the recognition that what makes taxa distinct are their apomorphies (or, instead, their unique sets of characters; apomorphies are usually single characters, not sets of characters). Species have diagnostic *aut*apomorphies—that is, they have unique constellations of characters—while higher taxa (clades) have *syn*apomorphies—shared constellations of characters, which group them together. One instance of this approach is found in the work of Donn Rosen: "a geographically constrained group of individuals with some unique apomorphous characters, is the unit of evolutionary significance" [Rosen 1978: 176, quoted in Wheeler and Platnick 2000: 55].

In Rosen's paper on Guatemalan fishes, he defined species in terms of individuals and populations: "a species is merely a population or group of populations defined by one or more apomorphous features; it is also the smallest natural aggregation of individuals with a specifiable genographic integrity that cannot be defined by any set of analytic techniques" [Rosen 1979: 277, quoted in Mayr 2000b: 99]. He then noted that this means that subspecies are, "by definition, unobservable and undefineable," since they have no such apomorphies. Subsequently, Nelson and Platnick proposed in passing that in their book on systematics and biogeography that they would treat species as "simply the smallest detected samples of self-perpetuating organisms that have unique sets of characters" [Nelson and Platnick 1981: 12, quoted in Wheeler and Platnick 2000: 56]. They, too, noted that this meant that diagnosable "subspecies" (groups identifiable by unique sets of characters) were thus species. This was not intended to be a complete definition but rather a description of their current practice (Nelson, personal communication). Even so, it was very influential on subsequent discussion. Independently, and almost contemporaneously, Eldredge and Cracraft [1980: 92, quoted in Wheeler and Platnick 2000: 55–56] defined a species as "a diagnosable cluster of individuals within which there is a parental pattern of ancestry and descent, beyond which there is not, and which exhibits a pattern of phylogenetic ancestry and descent among units of like kind." Cracraft

subsequently revised his formulation by removing mention of reproductive cohesion ("like kind") to read "the smallest diagnosable cluster
of individuals within which there is a pattern of ancestry and descent"
[Cracraft 1983: 170, quoted in Wheeler and Platnick 2000: 56, and see
discussion there].

Quentin (not Ward) Wheeler and Platnick base their definition on
this tradition as well [Wheeler 1999; Wheeler and Platnick 2000]. They
define species thus [p. 58]: "We define species as the smallest aggregation of (sexual) populations or (asexual) lineages diagnosable by a
unique combination of character traits. This concept represents a unit
concept. This concept is prior to a cladistic analysis" [p. 59], and so
unlike Nelson's earlier note that species are taxa like any other level of
a phylogenetic tree [Nelson 1989], they do not worry about apomorphies and homologies when recognizing species, only characters. Species
are found wherever characters are fixed and constant across all samples,
while traits may be variable [see their figure 5.1 on p. 58]. Their will-
ingness to handle and include asexual taxa within their species concept
marks them out from most other conceptions, and they bite the bullet
on recognizing clones of asexual lineages as species: "If the goal of distinguishing species is thereby to recognize the end-products of evolution,
should we seek to suppress naming large numbers of species where large
numbers of differentiated end-products exist?" [p. 59.].

Mishler considers a number of the diagnostic accounts to be phenet-
ically based, including Cracraft's, Platnick's, and Nixon and Wheeler's
(personal communication). Whether this is so (that is, whether they make
use of the Cartesian clustering of species in a state space of traits typical of phenetic practice), it is clear that they assume that species are phy-
logenetically speaking the terminal taxa on a tree. Diagnosis assumes that
the traits are specified before the tree is constructed.

Where Is the Taxon Level?

Operationally, phylogeneticists seem to have little operational difficulty
in identifying the level of species taxa in their cladograms, but there is a
problem that arises if the autapomorphic concept is taken too strictly.
Consider cladograms of haplotype groups. The authors have clearly already identified the species and are considering whether or not the haplotype data taken from, say, mitochondrial DNA support the claim that
these populations form subspecies. But on a strict autapomorphic concept, *each* of the haplotype groups should be considered a separate species

(in exactly the way Whately described a logical notion of *species* would do for dog breeds), unless other considerations, such as biogeography and interbreeding, are taken into account; and if they are, then the species concept alone is insufficient to delimit species. And so it appears that some sort of prior knowledge is required to specify at what level of a cladogram taxa begin and (for example) molecular lineages cease to be subspecific diagnostic criteria.

The claim that species are defined by constant characters, made by Wheeler and Platnick, is problematic. Either we already know what characters count as species defining, or we are unable to find a level of species (for there are constant characters for a great many higher-level groups, as well as lower-level groups, than the usual level at which species are identified). Either way, this is not a full concept of what makes a group a species.

OTHER SPECIES CONCEPTS

Ecological Species Concepts

Turesson's species concept did not get taken up widely, but ecological concepts have been proposed from time to time. Mayr himself ventured one in his 1982 history: "A species is a reproductive community of populations (reproductively isolated from others) that occupies a specific niche in nature" [1982: 273, italics in original]. He did this with neither preamble nor follow-up, and the requirement for the "niche" here seems to have been allowed to quietly drift away, as he does not insist on it elsewhere. A prior instance of another partial ecological concept occurs in Ghiselin, who referred to species as "[t]he most extensive units in the natural economy such that reproductive competition occurs among its parts" [Ghiselin 1974b: 38]. The competition here is for genetic resources, and comes in the context of a strong selectionist account of evolution at various levels [Ghiselin 1974a]. Ghiselin calls it the "hypermodern species concept" and says that species are economic entities analogous to firms.

Van Valen, in a paper that discussed the "odd" reproductive dynamics and evolution of American oaks (genus *Quercus*), proposed that in these plants at any rate a species was an ecological type. He offered a definition as "a vehicle for conceptual revision, . . . not a standing monolith": "A species is a lineage (or a closely related set of lineages) which occupies an adaptive zone minimally different from that of any other lineage in its range and which evolves separately from all lineages outside its range" [1976: 233]. This is self-consciously a mixture of the Mayr and Simpson definitions, and Van Valen justifies it in terms of evolution acting on phenotypes, controlled by "ecology and the constraints of individual development." It is therefore a definition founded on a particular view of evolution. He reprises Simpson's notion of a lineage and defines a population in genetic terms. The novel element here is the idea of the "adaptive zone," which he describes as "[s]ome part of the resource space together with whatever predation and parasitism occurs on the group considered" [p. 234]. *Quercus* species are often sympatric, and freely hybridize, yet they maintain their identity. He refers to these groups as "multispecies," as they exchange genes, but he does not require that they actually form viable hybrids that breed true thereafter. "Multispecies" is defined as a "set of broadly sympatric species that exchange genes in nature" [p. 235] and refers to the syngameon concept of Verne Grant, which, we have seen, is due to Poulton. Multispecies can occur without having component species as such, citing *Rubus*, *Crataegus*, and dandelions. The latter leads to the implication that asexuals are not to be considered species.

Elsewhere, Littlejohn [1981] has comprehensively reviewed reproductive isolation, and he makes some interesting comparisons between sexual reproduction and asexual isolation. He notes that asexuals (uniparental species; unlike many of his predecessors, Littlejohn does not exclude them from specieshood) must be seen as a "cluster or cloud of individuals representing an adaptive node or adaptive peak," citing Dobzhansky and G. E. Hutchinson [1968]. Hutchinson employs a "taxonomic space" model and treats species as clusters in that space, formed through adaptation. This is somewhat different to the phenetic concept of an OTU. For a start, he requires independence of the axes of the space and that they be adaptive characters. Asexuals, such as bdelloid rotifers, are just as good species as their close sexual relatives, the *Nebalia* class of crustaceans.[11] Similar points about the role of adaptation in delimiting species were previously made by entomologist R. S. Bigelow, who noted, "Reproductive isolation [in Mayr's 1963 definition—JSW] should be considered in terms of gene flow, and not in terms of interbreeding, since selection will inhibit gene flow between well-integrated gene pools despite interbreeding" [Bigelow 1965: 458].

More recently, the philosopher Kim Sterelny [1999] defended an "ecological mosaic" conception of species, based on a reworking of Dobzhansky's [1937a: 9–10] metaphor of species as occupying "adaptive peaks" in a Wrightean fitness landscape. Noting that species are typically *ecologically fractured* [p. 124], Sterelny concludes that most species are geological and ecological mosaics, and are not ecologically cohesive entities. He takes this to explain stasis in the duration of species, as interbreeding between populations within the species' range acts as an inhibitor, which he calls "Mayr's Brake," on adaptive change over all the entire species. He thinks that this reinforces an evolutionary concept of species.

Ecological species concepts are generally partial rather than all-encompassing, and they tend to act as adjuncts to more universal conceptions. It is clear that all ecological concepts rely heavily on the preeminent role of natural selection to maintain isolation between groups. Selection also plays a critical role in the definition and delimitation of asexual species, to which, among others, we now turn.

"Aberrant" Concepts

"Aberrant concept" here usually merely signifies that the concept is either not attended to much by zoologists or not thought by some biologists to be a "true" species concept.

Agamospecies Asexual taxa are *agamospecies*. The term was defined by Cain [1954: 98–106] as the end result of parthenogenesis in animals and apomixis in plants (which is secondary asexuality), but it is now well understood that most unisexual organisms are in fact neither [i.e., are primarily asexual, mostly bacteria or algae; Kondrashov 1994; Schloegel 1999; Taylor, Jacobson, and Fisher 1999]. Cain defined agamospecies as "those forms to which [the biological species concept] cannot apply because they have no true sexual reproduction" [p. 103].

This is clearly unsatisfactory. For a start, it is a privative definition—agamospecies are what are *not* something else (in this case, taxa comprised of sexual organisms). As we saw also with Fisher, asexuals are considered by Cain to be off the mainstream of evolution, something of a marginal occasional misfiring of the evolutionary process, since sexual recombination permits the more rapid acquisition and spread of favorable mutations, and asexual reproduction is subject to genetic load (the continuation of deleterious mutations). This is no longer the consensus.

So it has been questioned whether agamospecies are anything more than the morphological species concept applied to asexual organisms [noted by Sonneborn 1957: 283 of Dobzhansky 1937b, 1941; see the earlier discussion on Dobzhansky]. However, there is a notion that applies to asexuals that is, I believe, a more coherent way to understand asexuals and that avoids privative definition. Originally defined for viruses (which are mostly, but not always,[12] asexual), the concept of Manfred von Eigen [Eigen 1993a, 1993b; Stadler and Nuno 1994] is called *quasispecies* (from the Latin for "as if," *qua si*), and it applies equally to uniparental lineages in artificial life simulations as in biology [Wilke et al. 2001]. He notes that "[a] viral species . . . is actually a complex, self-perpetuating population of diverse, related entities that act as a whole" [Eigen 1993b: 32], and he defines the quasispecies as a "region in sequence space [which] can be visualized as a cloud with a center of gravity at the sequence from which all mutations arose. It is a self-sustaining population of sequences that reproduce themselves imperfectly but well enough to retain a collective identity over time" [p. 35].

The conception is based on the observation that in a cluster of genotypes of viruses, there will be a mean genotype (the wild type) maintained by selection for optimality for that particular environmental niche (figure 11). Eigen noted that it may eventuate that there is actually no single virus with that "wild-type" genotype. Clearly, this makes the agamospecies conception a kind of ecotypical notion, maintained by

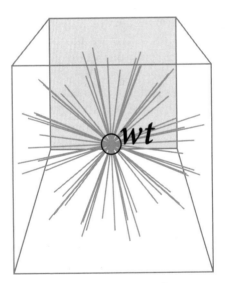

Figure 11 Eigen's conception of a quasispecies. Developed first to cover molecular and then viral species, Eigen's notion assumes that there is an "optimal wild-type" genome, which is here represented as a circle in an abstract "sequence space," divergences from which form a cloud of sequences constrained by selection. This applies to asexual organisms not otherwise constrained (e.g., by developmental constraints).

natural selection; in fact, one might say that it is the purely ecological aspect of the ecospecies conception. Similar observations were made by Hutchinson [1968: 184], but no technical name was applied by him at that time. I therefore propose that *agamospecies* and *quasispecies* be treated as synonyms and that the inferences made about quasispecies by Eigen be applied to agamospecies hereafter. [I elaborate and defend this approach in connection with microbial, mostly bacterial, organisms elsewhere; see Wilkins 2007a.] Bacteriologist Frederick Cohan has a number of papers on his *bacterial species concept* [Cohan 2001, 2002, 2006].

Nothospecies The almost exact conceptual opposite of the agamospecies/quasispecies conception is pteridophytologist W. H. Wagner's [1983] conception of a *nothospecies*. Effectively, this is a species formed from the hybridization of two sexual species (species formed by the usual method of cladogenesis of sexual species he refers to as *orthospecies*; personal communication). As Hull notes [Hull 1988c: 103, citing Verne Grant], "According to recent estimates, 47 percent of angiosperms and 95 percent of pteridophytes (ferns) are allopolyploids," and ferns and their allies continue to cause problems for isolation conceptions of species [Barrington, Haufler, and Werth 1989; Paris, Wagner, and Wagner 1989; Ramsey, Schemske, and Doyle 1998; Yatskievych and Moran 1989; Haufler 1996; Vogel et al. 1996; Wagner and Smith 1999]. The notion is so generally applicable in botany that the concept of nothospecies has been written into the botanical rules for nomenclature.

Compilospecies Harlan and de Wet [1963] proposed a concept of a species that "plunders" (Latin *compilo*) the genetic resources of another species through introgressive hybridization (where the hybrids preferentially interbreed with only one of the parental species, causing a gene flow from one to the other, one-way). It is therefore an asymmetrical version of the nothospecies concept. It applies to some plant species [Aguilar, Roselló, and Feliner 1999].

OTUs and Phenetics (Phenospecies) Phenetics developed from the work of Sokal and Sneath [1963; Sneath and Sokal 1973] who trace their attempt to produce a "natural" taxonomy back to Adanson, who used a multivariate classification scheme in *Familles des plantes* [Adanson 1763; Winsor 2004 provides a rebuttal for this claim]. The phenetic view, also called *numerical taxonomy* by virtue of the use of computers for the first time in systematics, basically involves a cluster analysis of continuously varying characters in a Cartesian space of n-dimensions (one dimension for each character or principal component) in the belief that species will fall out as clusters that agglomerate in different ways. While Sokal and Sneath [1963: 30f.] generally give priority to the biological species concept [but see Sokal and Crovello 1970], they note, "In the absence of data on breeding and in apomictic groups . . . , the species are based on the phenetic similarity between the individuals and on phenotypic gaps. These are assumed to be good indices of the genetic position, although they need not be. . . . In this book it [the term *species*] will be used in the sense of phenetic rank."

Species are not unique in the phenetic approach; they are operational taxonomic units (OTUs) just as above-species and within-species ranks are [pp. 121f.]. The phenetic view, though, is that species are not monotypic, and Sokal and Sneath define a term, based on Beckner's discussion [1959], *monothetic*, and its antonym *polythetic* [pp. 13–15] to describe groups based on a single uniform set of characters versus those which vary or have alternative characters; Wittgenstein's family resemblance predicate is explicitly adduced [p. 14].

Species Deniers: Pure "Nominalism" or Eliminativism

It is unfortunate that the alternative to the reality of species has been called "nominalism" by Mayr and others, because nominalism is a philosophical doctrine that asserts that *universals* are not real, and species are not held by many to be universal terms. Strictly speaking, the individuals that "species nominalism" considers real in opposition to species are individual organisms. However, in logic, a universal is not *comprised* of individuals, so much as individuals *instantiate* the class. Species *are* comprised of individuals under evolutionary accounts, and so they are real to the extent that their components are (except under ideal morphological views). It would be wrong to call these "nominalistic views" simply because they are founded on acts of naming. A species nominalism must be directed to the *category* or *rank* of species, and it must claim that there is no sense in which that categorical term has any application. I therefore prefer to refer to *species deniers* than to species nominalists.

Species deniers include Vrana and (Ward) Wheeler [1992], Pleijel [1999, 2000] and Hey [2001a, 2001b]. I have argued above that Darwin was not a species denier and that Buffon was only inconsistently one.

There are two varieties: *conventionalism* and *"replacementism,"* the latter involving replacing that term (*species*) with another [e.g., *deme*, Winsor 2000]—as Pleijel (LITUs) and Hey (evolutionary groups) do. Given the history of the term, there is no reason to do this, except to make it clear that only a particular sort or kind is being referred to, and history shows also that such attempts are always assimilated and subverted. *Species* keeps winning out.

Conventionalism: The Taxonomic Species Concept

> I object to the term "species concept," which I think is
> misleading. . . . A species in my opinion is a name given
> to a group of organisms for convenience, and indeed of
> necessity.
> **John B. S. Haldane [1956: 95]**

One major stream of thought, particularly among geneticists, with regard to species is what we might call the *conventional* account, although it is sometimes called the "nominalistic" or "cynical" concept [Kitcher 1984]. On this widely held account, species are "whatever a competent taxonomist chooses to call a species." The complete quotation from Charles Tate Regan [1926: 75], given by Julian Huxley [1942: 157; see also Ghiselin 1997: 118; Trewavas 1973] is "a species is a community, or a number of communities, whose distinctive morphological characters are, in the opinion of a competent systematist, sufficiently definite to entitle it, or them, to a specific name." Huxley goes on to note that the "difficulty with this definition lies in the term *competent*, which is what we have recently learnt to call the "operative" word. And experience teaches us that even competent systematists do not always agree as to the delimitation of species." Ghiselin notes that there are no rules for deciding whether a reproductive community is a species or a subspecies, and that one should wonder whose view to accept when the experts disagree; and such disagreement is common.

In fact, this view is not new and precedes Darwinian evolutionary theory, at least in Britain, for some time. Darwin himself was a member of the drafting panel that proposed just this standard for the new *Rules of Zoological Nomenclature* in 1842, the so-called Strickland Code [Strickland et al. 1843]. In so defining the basic taxon this way, the British Association for the Advancement of Science legislators "consciously and conspicuously distanced themselves from disputes over definitions of species" [McOuat 2001: 3] and, according to McOuat, established the naming rights of a species to be a delineation of what a competent naturalist was—basically, someone who was accepted by the naturalist community and, primarily, someone with a position at a museum, thus excluding names and species made by birdwatchers and gardeners. Darwin himself in several places made this sort of definition, particularly in his *Natural Selection*: "In the following pages I mean by species, those collections of individuals, which have been so designated by naturalists" [Darwin 1975: 98, cited in McOuat 2001: 4, n. 10].

It is interesting to note that Regan refers to "communities." This word typically has been used to apply to ecological communities—ecosystems [Taylor 1992]—but in this case and at this time it is more likely to refer to what we now call a "deme" [Winsor 2000] or a "Mendelian population." If this is so, then Regan is conflating two well-known ideas in our history here: that of a reproductive element and a diagnostic one. In short, Regan might very well have been putting forward an "operative" notion of the generative species conception we have so often encountered. A less biological version of this view, which he calls the "cynical species concept," is presented by Philip Kitcher: "Species are those groups of organisms which are recognized as species by competent taxonomists" [1984: 308]. Here the operationalist aspect of the concept is primary. Species are made by acts of recognition by experts. Whether or not it includes a strong element of the biological (that is, reproductively isolated) nature of species, conventionalism takes seriously Locke's claim that species are made for communication, and objections to the recent PhyloCode proposal (many by strict cladists, no less) are in part founded on the idea that higher taxa, at least, should be convenient, since they are artificial taxa, anyway.[13]

LITUs (Least Inclusive Taxonomic Units) Frederick Pleijel, a leading specialist on polychaete worms (bristleworms), has proposed doing away with the notion of species altogether [Pleijel 1999; Pleijel and Rouse 2000]. Instead, he proposes to replace it with the neutral term *least inclusive taxonomic unit*, or LITU (in homage to the OTU of phenetics). Pleijel and Rouse's [2000: 629] "definition" of the LITU is "named monophyletic groups which are identified by unique shared similarities (apomorphies). . .which are at present not further subdivided. . . . Identification of taxa as LITUs are statements about the current state of knowledge (or lack thereof) without implying that they have no internal nested structure; we simply do not know if a given LITU consists of several monophyletic groups or not." However, as the historical reviews cited earlier make abundantly clear, the notion of a least inclusive taxonomic unit historically *is* a species in both logic and biology. What Pleijel has done here is rediscover the past way of looking at things.

Species Concepts in Paleontology (Paleospecies)

The problem of applying any concept of species in paleontology has been long understood and has spurred a number of discussions [Simpson 1943; Sylvester-Bradley 1956; Schopf 1972; Smith 1994]. The difficulty lies in the way the data are presented. In neontology (the study of living organisms), the behaviors of organisms, both sexual and ecological, can be observed in some detail and repeated if the initial observations are inadequate. Under some circumstances, the organisms can be experimentally mated. Molecular evidence, in particular that of DNA, can be harvested and assayed, meaning that where a traditional taxonomist might have used at most around forty characters, the molecular systematist has more like forty thousand.[14] Also, polytypy can be investigated in extant species, enabling the investigator to delineate the populations and subspecific variants and to tell whether, on the concept used, these still are included in the specific group.

Not so the paleontologist. The information in that case is usually restricted to several individual specimens (except when the bulk of the population leaves fossils, as in the case of silaceous forams, whose shells are fossilized in sediments on the seafloor, and which when they are found provide information about distribution, variety in populations, and changes over time). The famous *Tyrannosaurus rex*, for example, is known from around twenty specimens, not all complete.[15] Many hominid species are known from a single individual. In cases where the ancestors of a lineage lived in conditions unconducive to fossilization—the study of which is known as *taphonomy*[16]—we will have entire series of species unknown to science. For example, few ape fossils have been found for those species that lived in forest and jungle environments, where decomposition in the acidic soils, scavengers, and plants will dispose of the carcass relatively quickly.

So paleontologists rely almost exclusively on morphological data. This means that there is pressure on them to lump stratigraphic specimens together (since fossil taxa are often used as stratigraphic markers by geologists) if there is some subjectively acceptable similarity between specimens.[17] In the rare cases when many specimens are found, problems due to variation can cause taxonomic problems. If many locales are involved—that is, if the specimens are allopatric—it is unclear whether they form a single species or allopatrically isolated but closely related sister species. The problem is critical in the case of *Homo erectus*, which has a range from southern Africa to eastern and southeastern Asia. If all that remains is skeletal information, can we really be sure if these are the same biospecies, for example? The now-amalgamated species *Canis lupus* includes morphs like the timber wolf, the Pekinese pug, and the Great Dane, all interfertile or fertile along a series of intermediate forms. Any paleontologist would split them into distinct taxa on morphological grounds alone. In the case of many birds, on the other hand, skeletal changes between good species are minimal, and only behavioral and ecological differences will tell them apart.

Chronospecies (Successional Species) A problem that used to be common in discussions of paleontology and species concepts was that of chronospecies, where speciation occurs over time such that at the starting and end points of a time series, the morphs are different species. This was discussed by Simpson, for example, in his 1944 book on evolutionary tempo and mode. However, the notion of chronospeciation is no longer thought by many to be an operational one; for a start, a number of specialists think that for animal species, such changes do not occur without lineage differentiation—in short, anagenesis is accompanied by cladogenesis—and that in most cases species remain largely unchanged after their initial period of adaptation until they go extinct [Eldredge and Gould 1972; Eldredge 1985; Eldredge et al. 1997; Gould 2002].

Chronospecies are formed, according to the author who proposed the notion [George 1956: 129], when a lineage changes sufficiently to be given a new name. In this respect, they are a temporal version of the taxonomic species concept, or even an interpretation of the paleospecies concept [Cain 1954: 106f.; Simpson 1961: 166]. A chronospecies is effectively an arbitrary division of a gradually evolving lineage [Eldredge 1989: 98] and seems to have been come to prominence as the "species concept" of "phyletic gradualism," the target of criticism of the theory of punctuated equilibrium [Eldredge and Gould 1972; Gould and Eldredge 1977], although it is not often used in the interim, so far as I can tell.[18]

HISTORICAL SUMMARY
AND CONCLUSIONS

It is time to sum up the major claims of this book, and they are many. For a start, we have considered several historical and several philosophical claims, each in the light of each other. The biological conclusions are not mine to draw, but I can opine—and do.

From Aristotle through to the end of the Middle Ages and the Renaissance, the notion of species has not remained static; Aristotle's conventions and notions have been modified. Mostly, they were modified by the neo-Platonists and especially by Porphyry, who made Aristotle's top-down classification scheme dichotomous after the manner of Plato. Aristotle had opposed "privative" classification, classifying in terms of what things are *not*, although the grander schemes of later writers up to Lamarck had little problem with this and happily classified groups such as Invertebrata.

In the medieval scheme, the notion of genus and species did not involve fixed ranks; a species might in turn be a genus on its own. The only "absolute" ranks were the *summa genera*, which represented in the Aristotelian tradition the universal categories (topics, or *topoi*) from which all things were to be divided. The nominalist issue of whether these general terms were merely aspects of mental categories or were real was alive and active well into the scientific period.

We find in the Epicureans a *generative conception of species* as early as the fourth century BCE, and this recurs throughout the remaining discussions until the biological tradition begins in the seventeenth century. Essences, on the whole, were not themselves necessary and sufficient criteria for membership in species, and almost all writers admitted that there were deviations from the type. The classical scheme was, however, almost always based on a top-down classification with a large admission of apriorism, until the collapse of the Universal Language Project and the rise of corpuscular philosophy, which rendered species secondary qualities or unknowable.

The Great Chain of Being meant that, depending on what emphasis was given to the principle of plentitude and the principle of continuity, species were arbitrary divisions in a plenum, sometimes logical, sometimes substantial and actual. The specific nature of a member of a species was thought by Cusa to be a contraction of the essence in that individual, but the ultimate reality, according to Cusa, is the individual. A continuing battle between nominalists and realists (idealists, in modern terms) meant that there was a field of alternate opinions. Variation is recognized to be a fact within taxa from the fifteenth century onward. Some, such as Ficino, held that there was a species that was most representative of a genus, since other species within the genus could play on variations of the generic theme, as it were.

With Bacon we move from an immediate generalization of the universals from observed instances, and a subsequent top-down division of things, to the inductive construction of increasingly broader generalizations. He, too, allows for deviations and variations in species and other taxa. Locke proposes not only that there is biological variation but that species (sorts) themselves are conventional names we use to communicate easily. Nevertheless, he did allow for a real essence, only it is one that cannot be defined or even discovered. Leibniz was more optimistic; Kant, even more pessimistic.

Kant rejects Leibniz's view of the law of completion (principle of plentitude) and argues that nature is discretely divided; he, too, uses a generative conception of species. Although the traditional logic survives until the institution of set theory in the nineteenth century, even Mill is able to twist it to serve biological realities. Both the logical tradition before Darwin and that after, in his own country, allow for a difference between essentialist logical species and typological biological ones.

There is everywhere a remarkable lack of the sort of essentialism that Mayr and others believe permeates this period and its philosophy. While we see typology, when it comes to dealing with biological organisms, most of the time there is no insistence on essences, and sometimes there is an explicit exemption for biological species of any knowable essence.

The major break with the classical notions of species from the tradition of "universal taxonomy," as I have called it, has been allied with the Baconian insistence that instead of beginning classification with the universals, species of living things are found by a process of ascending abstraction and generalization. Species have been understood to be propagative forms, which I have here termed the *generative conception*, and in various degrees and emphasis, this remained the basic conception until the modern era. For this reason, the focus has been on seed, the fructative apparatus, and the reproductive behaviors, according to then-current views on generation [Gasking 1967].[1] The reason for a focus on species as units of biology derived from the medieval tradition and the neo-Platonic revival of the seventeenth century, but this was richer than the Received View supposes.

Many conceptions of species depended on, or were confounded by, the Great Chain. In one respect, a species was whatever was lowest in generality, but in another, species were conventions or artificial divisions due to the fact that not all variant forms are to be found in a locality. Plenitudinous views expected that intermediate forms would be found, and, right into the nineteenth century with Macleay and Swainson, some held that while species might be discrete, there would be a continuity of form itself. There was often a species regarded as most like the vanilla generic characters, the *type species*, which relied on the Great Chain conception. There was disagreement about the level of organization, too. Buffon thought that local forms were variants of the true species, the *première souche*, while Linnaeans tended to name any persistent variant form as a species. Bonnet and the early Buffon thought that specific forms were mere abstractions, a nominalist view that gradation imposed also on Lamarck. Even so, all these writers made use of a generative conception. Species may not exist, but if they did, they were the result of heredity on the reproduction of forms.

Species realists in the later biological tradition tended to be fixists, such as Cuvier and Agassiz, but, of course, such a contrast was not possible until the possibility of transmutation over time was mooted. In the period before fixism was introduced, typology was widespread, but species were at best only mutable, not transmutable, excepting the case of spontaneous generation, in which species could sequentially change in outward form, rather like the metamorphosis of insects. The process by which species came to be in the first instance, however, was irrelevant to how they were maintained, and a generative notion was used by realists as well, especially those who stressed form as an identifying or diagnostic factor. Most of them differed from the transmutationists as to whether the formal diagnosis was a mark of the real essence; fixists thought that the generative process was the reason for the form, transmutationists that the generative process (Lamarck's *feu éthéré*, for example) was the reason only for the form in a given time or place or conditions. In short, for fixists, the form gave the essence, while for transmutationists, the "essence" meant that the form would not remain constant. Some like Buffon and A.-P. de Candolle bridged the two camps, declaring that the real essence of a species was modified to a certain degree by local conditions and hybridization.

In the period before the *Origin*, in Britain and presumably elsewhere, species were declared to be the "property," as it were, of competent and recognized taxonomists, particularly in museums, as McOuat has discussed. A number of experts doubted the permanence of species, so that by the midcentury, several writers had made the suggestion in reputable forums that species did change.

Darwin's own views changed. At first he was quite comfortable with the standard generative notion, and made notes on the idea that it was either the physical impossibility of interbreeding or the "repugnance" of species to interbreed in the wild. Diagnostic issues followed from the facts of the matter (the nominal essence did not give the real essence, in other words). But early on he started to toy with the notion that species were meaningless and conventional, and at the time of the *Origin* and shortly after, he seems to have taken this as his position—species are the outcomes of the evolutionary process acting on varieties and are not real entities themselves. It doesn't seem to have made much impact on his own systematic practice, as it was not to do for anyone else, either, for some time to come. The *rank* of species is arbitrary in the *Origin*. However, species could "do" things—they could even compete with each other.

Darwin first thought that geographic isolation formed species most of the time but by the *Origin* had shifted to the view that selection against intermediates and hybrids was the major force. The geographic view of speciation, though, was to become the majority position in opposition to Darwin except for a few of his most devoted followers. However, while he may have been a conventionalist in his view of identification, he was not a nominalist, unlike several of his followers. He was a pluralist as to the degree of difference between, and causes of, species. It is significant that he thought that sexual constitutions were what caused the isolation of species, not form or adaptive traits. Darwin's evolving views are significant, as his is perhaps the first and the most complete attempt to deal with the implications of the transmutation of species.

The adaptationists such as Wallace, Romanes, and several others, however, were monists. Species were formed solely by selection. Weismann is an interesting exception here—he appears to have allowed for stochastic causes as well as deterministic selectionist causes of species. Poulton's essay, discussing these turn-of-the-twentieth-century arguments, effectively set the stage for the modern debates by introducing the notion of a genetic population as the boundary of species. Several contemporary writers such as Lotsy, Jordan, and Turesson elaborated on that theme under the recent introduction of Mendelian genetics, with Turesson reintroducing the older notion of ecological habitat affecting form (albeit in a much more limited way). Several authors treated asexuals as a different kind of entity to species; Fisher even thinks asexuality is not a real phenomenon. This will become significant later in the century.

One thing that is obvious, even at this early stage in the debates over the "species problem," is that there is a distinction to be made between "universalist" species concepts, which are intended to apply to all organisms, especially in regard to speciation mechanisms after Darwin, and those proposing more limited notions, which apply only to some sorts of organisms. In the twentieth century, most conceptions of species are universalist conceptions, and the properties that mark them out are those derived from the proponent's preferred mechanisms of speciation. Biospecies are formed through the acquisition of reproductive isolating mechanisms in allopatry or sympatry according to the preference of the author. Saltationists claim that all species are formed through macromutations. Punctuated equilibrium theory later treats speciation as a rapid process followed by a period of stasis, and so species are universally delimited by this "sudden" event, thus finding the individualism thesis and the evolutionary conception of species congenial, and so on. Only a few people proposed limited conceptions, such as ecological conceptions of agamospecies, which had no general application.

$$\cdot \quad \cdot \quad \cdot$$

So let us list the conclusions of this book in summary form.

Genus and *species* and their cognates in Greek and other languages are vernacular terms, logical terms, and biological terms. It is important to read each text in the right context and understand that when classical authors, including Aristotle, uses these terms (or their Greek equivalents), they may be using them in a vernacular or technical logical sense. The biological notion of *species* did not develop until the end of the sixteenth and through the seventeenth centuries CE. Consequently, it is anachronistic to read the classical writers as making claims about biological species.

Throughout the history of biological thought, *species* has always been thought to mean the generation of similar form. That is, a living kind or sort is that which has a generative power to make more instances of itself. The generative conception of species was the common view from the Greeks to the beginnings of Mendelian genetics around 1900. Prior to this, there had been a "species question"—that is, the question of the origination of new species. After this, there was a "species problem," in which various attempts were made to identify the genetic substructure of species. The changeover dates to Poulton's essay in 1903.

There has been no morphological species tradition as such, apart from the use of morphology to *identify* species, and this continues today. Moreover, idealism and morphology are distinct programs—not all morphologists are idealists (and, hence, not all who rely on morphology are Platonists).

Species have always been understood to involve deviation from the type, from Aristotle to the modern era.

Type and *essence* are distinct ideas in biological history. Types can have variation, while essences cannot. Systematics has always used the type concept and a "method of exemplars," but rarely has it used essences as anything but a useful set of diagnostic keys or as an aid to identification. *Essence* can be understood as a *formal* notion—one used in description, definition, or identification—or it can be understood to be a *material* notion—in which the essence *causes* the thing to be what it is. Many biologists followed Locke in supposing that the Nominal (formal) Essence was not the Real (material) Essence. There is no substantial material essentialism in natural history or biology until *after* Darwin, if then. What there is, is probably a reaction to Haeckel and the "evolutionists" who followed Schopenhauer and Nietzsche, and it largely relies on the revival of Thomism after Vatican I. There is a minor tradition, scientifically speaking, of neo-Thomist-inspired scientific material essentialism from around 1870 to the end of the 1960s.

The synthesis Darwinians generated and promoted a history of *species* before Darwin as essences based on a misreading of the (mostly logical and metaphysical) sources and applying them incorrectly to biological cases. Scientists often use history as part of a program to promote their current scientific views, either by demonizing their opponents in proxy or by demonstrating that they are the culmination of a progressive historical process of discovery. This is not restricted to either side of any debate. Whiggism is rife in textbook histories.

The overall problem of species derives from its neo-Platonic history as a top-down category of the logic of classification [Boodin 1943]. Modern taxonomy works in the opposing direction—beginning with the organisms, the individuals in the medieval system, and thence to lineages, populations, and then species. Species in biology are the result of inductively generalizing from individuals, rather than dividing general conceptions into subaltern genera to reach the infimae species. We still desire to treat *species* as a natural kind of term, and hence to find essential features that define all and only those taxa. Between-species synapomorphies are not like this; all they have in common is that they keep lineages distinct (either causally or cognitively, the ontological and the epistemic sides of this issue). They necessitate a bottom-up classification logic or, perhaps better, an *in media res* logic [cf. Ghiselin 1997: 182ff.]. The reason I have made such play with cladistic conceptions of classification in this work is that cladistic classification is—depending on how you interpret the matter—either a prolegomenon to induction, or an act of induction in itself. Bottom-up classification involves projectible inductive inferences and predicates, and "being a species" is one of these predicates. We predicate of some group that it is a species, and we mean by that, that it is held distinct from other groups; this is all that the biological taxon concept has in common with the medieval conception of classificatory categories. We need to resist the tendency to fall back into the older way of thinking about classification. There is no universal grammar or language of nature. John Ray was right, and John Wilkins (the *other* John Wilkins) was wrong.

One thing that ought to be clear from this book so far is that the standard stories and assumptions from the architects of the modern synthesis are often simply incorrect. What implications might the loss of the essentialist myth have for scientific research? We might no longer see it needful to deny there is a human nature, for example, while remaining true to the understanding that there is no human *essence*. This could affect our approach to such topics as evolutionary psychology. We might begin to see that formal considerations and biological considerations do not immediately intertranslate and so defuse a good many arguments about classification. We might see that species can be real phenomena in, say, a local ecosystem, without requiring them to play the same explanatory role in every ecotype. We might stop trying to overgeneralize species concept(ion)s or speciation mechanisms to all species. This would reduce the heat in a number of biological forums. And we might just value conceptual clarity and stop trying to employ the dead in support of modern views, while not overvaluing modern views at the expense of a strawman of the past. Perhaps this will help biologists appreciate older work without the polemics and caricatures currently in use. And Suidae may evolve feathered forelimbs for locomotion.

Notes

PROLOGUE

1. There is a tendency in the philosophy of science to refer to an older viewpoint or school of thought as the "Received View." The Semantic Conception of Theories presented by Suppe [1977] is contraposed to a Received View. Hull himself contrasts his evolutionary view to the Received View of Popper and Hanson [Hull 1988c: 484]. This tactic does run the risk of assuming a global hegemony of views against the "radical" nature of one's own, but until recently this view of species has indeed universally been received even by those who *want* to defend essentialism [such as Ruse 1987, 1998]. Amundson [2005] calls a similar tradition the "Synthetic Historiography," in a broader context. Winsor [2003], independently, also called something similar to this the "received view." Another more restricted term is "The Essentialist Story" [Levit and Meister 2006].

2. In fact, Joseph had not made the misinterpretation claimed but had been read wrongly by Cain and Hull, as we shall see.

3. Dominique-Alexandre Godron, a French botanist (1807–1880). According to Mayr [1982: 649], his major interest was in the nature of introduced species and the effects of cross-breeding [Godron 1854].

4. Hull, personal communication, 2002. As Hull's ideas have been so influential, it should be noted that also at about this time he published [Hull 1967] a brief history in which he went into more detail and that cites Joseph in note 3. In it, he notes that the observation that things do not always breed true was made by Aristotle and Theophrastus, and that therefore species *do* allow variation and divergence from the type. Even so, Hull still skips from Aristotle to Cesalpino, with only a short allusion to the Great Chain of Being, and while he refers to neo-Platonism, it is the neo-Platonism of the late eighteenth, not the classical and Renaissance, centuries.

5. Which I do not cover here. Some secondary sources indicate that not much happened in the Islamic high period with respect to biology, but given my experience with commentators on the Western sources, I tend to doubt it. Somebody with access to the material might like to investigate this period. One possibly important thinker is the ninth-century Persian writer, Al-Jāḥiẓ, who compiled an extensive *Book of Animals (Kitab al-Hayawan)*, and another the twelfth-century Andalusian author Al-Marwazī, who wrote *The Natures of Animals*.

6. See also Amundson [1998]. Challenges to the typological/essentialist account begin, so far as I can see, with Farber, Platnick and Nelson, and Panchen in the period from 1970 to 1985 (references cited later).

7. I am indebted to Kim Sterelny for this point.

8. Among others. Nicolai Hartmann appears to have been a philosophical influence, too [Hennig 1966: 22].

9. Despite its title, Ghiselin's paper "On Psychologism in the Logic of Taxonomic Controversies" [1966] does not address the psychological origins of taxa but the introduction into taxonomy of epistemological conventionalism.

10. I am influenced by David Hull's evolutionary conception of science [Hull 1973a, 1983a, 1984a, 1984b, 1984c, 1988a, 1988b, 1988c, 1990, 1992] and have tried to present my own evolutionary account of science and culture in this vein before [Wilkins 1998, 2001, 2002]. If there is a difference now, it is that I no longer expect ideas necessarily to be subjected to selection—that is, memes, like genes, drift [Wilkins 1999]. However, on Hull's view of science combined with Sergey Gavrilets's recent work on adaptive landscapes, these two processes need not be in opposition [Wilkins 2008].

THE CLASSICAL ERA: SCIENCE BY DIVISION

1. This account owes its outline and many details to Nelson and Platnick [1981] and Panchen [1992].

2. "History" is, in this older sense, the Greek word *historia*, which means an "inquiry or investigation," which derives from the title, or rather the opening words, of Herodotus's *History*. Later it comes to mean "knowledge" or "learning."

3. An excellent treatment of many of the themes discussed in this section, and a good aid to understanding the historical contexts can be found in Mary Slaughter's wonderful book of 1982; a broader and more liberal treatment, but one that suffers from overtheorizing, is Michel Foucault's *The Order of Things (Les Mots et les choses)* [1970], particularly chapter 5. Much of the source material and a detailed discussion of the *logica combinatorialis* can be found in Rossi's [2000] treatment relating mnemotechnical *(ars memoria)* traditions and the Universal Language Project.

4. Some believe that Plato was a Pythagorean who broke the mystery boundaries of the religion, and at least one person known to me argues that some of Plato's work is a guarded expression of Pythagorean mysteries regarding the ratio, or *logos* of number.

5. A good technical introduction to the received opinion on Plato's theory of Forms is provided by Windelband [1900, §35]. For a more updated treatment, see the introduction to Matthews [1972] or chapter 1 of Oldroyd [1986].

6. Nichols translates it as "to cut apart by forms, according to where the joints have naturally grown, and not to endeavor to shatter any part, in the manner of a bad butcher" [Plato 1998].

7. Quoted from Francis Darwin's *Life and Letters*, volume 2, p. 427, by Depew and Weber [1995: 43].

8. Using G. R. G. Mure's older translation for emphasis of the terms *genus* and *species*. A more recent translation [Barnes 1984] is less clear, using terms such as *atomic, primitive,* and *simple* in preference to *infimae species*. As an interpretation of the intent of Aristotle himself, the Barnes edition is probably better (although I have my doubts—it imports modern logical notions and so obscures Aristotle's preoccupations; but then, the older translations accreted much Scholastic and later connotations), but Aristotle has been mediated to the modern (postmedieval) tradition via the sorts of interpretation embedded in the Mure translation.

9. "[Classification] through class and difference." More fully, as *per genus proximum et differentiam specificam*—"through nearest class and specific difference" [Joseph 1916: 112]. Thanks to Tom Stevenson for the translation.

10. In other translations, such as W. D. Ross's [in McKeon 1941], the phrase is "the kinds of differentiae . . . characterized by the more and the less."

11. Balme says that in *Parts of Animals* 1.2–4 [Balme 1987b: 19]:

> Aristotle concludes that diairesis can grasp the form if it is not used dichotomously as Plato used it but by applying all the relevant differentiae to the genus simultaneously; after that he explains the ways in which animal features should be compared so as to set up differentiae—by analogy between kinds (genē), by the more-or-less as between forms comprised within a kind (eidē).

12. A point made by Hugo de Vries in 1904 [in de Vries 1912], intriguingly, although one might dispute his detailed claims about the genus being natural and not the species in Linnaeus's works. See the discussion on the medieval debates over the predicables.

13. The Perseus edition of the Latin text of Mayhoff numbers chapters differently from the English translation of Bostock. The Latin numbering is used here. See http://perseus.uchicago.edu/cgi-bin/ptext?doc=Perseus%3Atext%3 A1999.02.0137&query=head%3D%231, accessed May 5, 2005.

14. From the section "On Rhetoric" in Book I, translation by H. E. Wedeck [Runes 1962: 211–212].

15. As noted, Joseph's book has been implicated in the adoption by Cain and Hull of the notion that preevolutionary species are timeless and static entities defined by their essences. It should be noted that Joseph is presenting a formal account of the pre-set theoretic logics from Aristotle until his day, and he does, in several places [pp. 53n., 92–96], note the differences between Aristotle and Porphyry, but it would not have been obvious to anyone not familiar with the technical aspects of the medieval commentaries on Porphyry. Joseph's book is actually a very good late example of the treatment of logic in the Aristotelian

tradition, surviving into the post-Darwinian era but aware of it [pp. 473–475]. Hull [1967: 310, n. 3] cites a passage from the first edition of 1905 in which Joseph notes that the evolution of species is not thought through by biologists in the logical sense, but I cannot find it in the second edition.

16. The passage is translated from *Anicii Manlii Severini Boethii In Isagogen Porphyrii commenta* [2nd ed., Book I, ca. 10–11, ed. Samuel Brandt, p. 159, l. 3–p. 167, l. 20]. Translation by Paul Spade, unpublished, used with permission. Barnes's direct translation of Porphyry [Porphyry and Barnes 2003: 3, ll. 10–15] reads, "For example, about genera and species—whether they subsist, whether they actually depend on bare thoughts alone, whether if they actually subsist they are bodies or incorporeal and whether they are separable or are in perceptible items and subsist about them—these matters I shall decline to discuss, such a subject being very deep and demanding another and a larger investigation."

17. And interestingly, until late in the seventeenth century, "Realism" denoted a realism about *ideas*; what we would consider idealism today [Blackmore 1979].

18. Barnes agrees [Porphyry and Barnes 2003: 312–317], with more warrant than I have, that the traditional claim of Stoic influence has no basis in fact and conjectures [pp. 356–358] that there are many "Epicurean touches" in Porphyry. Preus [2002] notes that Plotinus, Porphyry's teacher, had made some passing comments on form *(eidos)* in the *Enneads* (V.9.6), which gives an Epicurean-style generative account of species, in which *logoi* are the generative powers "in the seed" and of every part of an organism. He says, "Some call this power in the seeds 'nature,' which was driven thence from those prior to it, as light from fire, and it turns and enforms the matter, not relying on the help of those much-discussed mechanisms (levers), but by imparting the *logoi*" [Preus's translation, p. 46].

THE MEDIEVAL BRIDGE

1. Amusingly, Walter Raleigh, in his 1614 *History of the World*, denies that the hyena is a true species, according to T. H. White, the translator of *The Book of Beasts* [p. 31 n.], because it was a hybrid form of cat and dog, and so would not have been on Noah's Ark.

2. But see Abulafia's biography [1988: 252ff.] for a dissenting account of how Norman Frederick's court really was, defending its sophistication and culture.

3. Benziger Bros. edition, 1947, translated by Fathers of the English Dominican Province, accessed at https://www.greatestbooks.org/studentlibrary/churchdoctors/aquinas/summa/index.html on July 3, 2005.

SPECIES AND THE BIRTH OF MODERN SCIENCE

1. This brings to mind Donne's much later use of the phrase "the world's contracted thus" in his poem "The Sunne Rising" (c. 1605). Donne was a one for lamenting the loss of the older medieval categories of thought, famously complaining in his "Anatomy of the World" (1611) that all coherence was gone

with the loss of Aristotelian physics and Ptolemaic astronomy [Kuhn 1959: 194].

2. Ong's book is singularly bad tempered but erudite. The author, a Jesuit and a neo-Thomist, is attacking the vogue Ramus's anti-Aristotelian views were having at the time. Ong's evaluation of Ramus is not followed here, even if his interpretation is.

3. A good summary, with excellent reproductions of woodcuts of the period, is Pavord's book *The Naming of Names* [2005], which also includes a nice discussion of the Arabic contribution to the herbalist tradition.

4. Arber herself, a botanist in the ideal morphology tradition, is a species "nominalist," writing:

> The progression from the vague concepts of the early writers to the sharp definition of genera and species to which we are now accustomed, has been in some ways a doubtful blessing. There is to-day, as a recent writer has pointed out, a tendency to treat these units as if they possessed concrete reality, whereas they are merely convenient abstractions, which make it easier for the human mind to cope with the endless multiplicity of living things. [p. 168f.]

The "recent writer" may be G. Senn [1925], whom she cites in her Appendix III [p. 306].

5. Arber [1938: 142f.] notes that Linnaeus's personal copy of *De plantis* is heavily annotated. She comments that "Cesalpino's strength lay in the fact that he approached his subject with a trained mind; he had learned the lesson which Greek thought had then, and has now, to offer to the scientific worker—the lesson of how to think."

6. "How like us is that very ugly beast, the monkey"—Ennius (239–169? BCE) as quoted in Cicero's "On the Nature of the Gods." Thanks to Tom Scharle for the reference and translation. *Simia* can mean "ape" or "monkey" and is most probably derived from *simulis*, "similarity" (to human beings).

7. The provenance of the saying Ray adopts here—*natura non facit saltum*—is interesting. Usually associated with Linnaeus, it can be found also in Leibniz and even, in a form, in Albertus Magnus: "nature does not make [animal] kinds separate without making something intermediate between them, for nature does not pass from extreme to extreme *nisi per medium*" [quoted in Lovejoy 1936: 79]. The idea can be traced back to the views of Plotinus and Porphyry, and probably also to the Gnostic idea of emanation. Ray also used similar phrases: *Natura nihil facit frustra* (Nature makes nothing in vain) and *Natura non abundant in superfluis, nec deficit in necessarius* (Nature abounds not in what is superfluous, neither is [it] deficient in necessaries)—in the *Wisdom of God* [quoted in Cain 1999a: 233]. This saying, adopted much later by Darwin, is also found in a similar form in Leibniz's *New Essays*: "In nature everything happens by degrees, and nothing by jumps" [Leibniz 1996, Book IV, chapter 16, p. 473]. It is an expression of the Great Chain of Being.

8. Possibly a title better held by Robert Boyle (M. P. Winsor, personal communication).

9. I am informed by Staffan Müller-Wille (personal communication) that Linnaeus, being from a relatively poor district of Sweden, Småland—known (presumably by an Englishman) as the "Scotland of Sweden"—was taught from

old standard textbooks, not out of the neo-Platonists early or late, as far as is recorded [see also Frängsmyr 1983; Koerner 1999; Goerke 1973]. According to Hagberg [1952: 44ff.], Linnaeus was greatly influenced by Aristotle's *Historia Animalium* as a young student.

10. According to the *Oxford English Dictionary, phylum* is a term first coined by Cuvier, in *Regne Animal* [1812], to cover his four *embranchements*, later adopted and made popular by Ernst Haeckel. *Family* is most probably Adanson's term [Judd et al. 1999: 40]. The Strickland Code of 1842 [Strickland et al. 1843: 119] mentions "families," noting they ought to be ended in -*idea*, and this implies families were in common use by that time. It also allows subfamilies. Mayr, Linsley, and Usinger [1953: 272] give the introduction of "family" to Latrielle in 1796, but they do not give any information regarding *phylum*. In botany, *phylum* is not used, and instead the rank is *division*, probably introduced in Alphonse de Candolle's 1867 Rules submitted to the Paris meeting that year of the International Botanical Congress. The present International Code of Zoological Nomenclature does not regulate higher taxon ranks above *superfamily* [Winston 1999: 32], so *phylum* is in effect an informal rank. Recent attempts to revise the rank of *kingdom* and add *empire*, or *domain* [Woese 1998; Syvanen and Kado 1998; Margulis and Schwartz 1998; Williams and Embley 1996; Baldauf, Palmer, and Doolittle 1996] are thus legitimated by tradition even if not yet widely accepted.

11. See Broberg [1983] for a comprehensive account of Linnaeus's treatment of humans in relation to apes and the reaction he received from the religious, both naturalists and theologians.

12. Some have held that *sapiens* should be read as "the knowing man" [Broberg 1983: 176]. The classical meaning of *sapiens* is "wise or discerning" or "sage," and it is used this way in Ovid, for example. Linnaeus, who quotes the Oracle's advice "know thyself" *(Nosce te ipsum)* on which Socrates based his investigations as the definition of *Homo sapiens*, may have been alluding to this. However, I still think *wise* is a better translation. Thanks to Polly Winsor for pointing this out.

13. In modern practice, the genus name is always capitalized, the species epithet is always lowercase, and both are always italicized. Other taxonomic ranks are capitalized but not italicized.

14. According to Mayr [1982: 870n.], the term *taxon* was proposed in 1926 by Meyer-Abich [see also Lam 1957, who discusses the term in more detail, noting that it is a nomenclatural term for a phylogenetic group]. Hence, in this context it is an anachronism. Stafleu notes that Linnaeus's own general term for taxa was *phalanx* but that it did not catch on.

15. He called them naturalists, of course, because the *term* biology was not coined until the end of the eighteenth century.

16. Full citations and page numbers to be found in Eddy [1994: 646n.].

17. The French is [Adanson 1763: clxviiij]:

Définition de l'Espèce.

Ainsi, quoiqu il soit très-dificil, pour ne pas dire impossible, de donor une définition absolue & générale d'aucun objet de l'Hist. nat. on pouroit dire assez exactemant qu'il existe autaunt d'Espèces, qu'il i a d'Individus diférans entreaux, d'une ou de plusieurs diférances quelkonkes, constantes ou non, pourvo qu'eles soient très-sensibles, & tirées des parties ou qualités où ces diférances paroissent plus naturelement placées, selon le génie ou les moeurs propres à chaque Famille.

18. The Latin of the definition is [Jussieu 1964: xxxvij] "in unam speciem colligenda sunt vegetantia seu individua omnibus suis partibus simillima & continuatâ generationum serie semper conformia."

19. *Gesammelte Werke* II: 429. Translation by Mark Fisher, used with permission. A version of this passage can be found also in Dobzhansky [1962: 93, from which I found this passage]. Kant has rightly identified the reason for the Linnaean system, which was at the time gaining ground among naturalists.

Im Thierreiche gründet sich die Natureintheilung in Gattungen und Arten auf das gemeinschaftliche Gesetz der Fortpflanzung, und die Einheit der Gattungen ist nichts anders, als die Einheit der zeugenden Kraft, welche für eine gewisse Mannigfaltigkeit von Thieren durchgängig geltend ist. Daher muß die Büffonsche Regel, daß Thiere, die mit einander fruchtbare Jungen erzeugen, (von welcher Verschiedenheit der Gestalt sie auch sein mögen) doch zu einer und derselben physischen Gattung gehören, eigentlich nur als die Definition einer Naturgattung der Thiere überhaupt zum Unterschiede von allen Schulgattungen derselben angesehen werden. Die Schuleintheilung geht auf Klassen, welche nach Ähnlichkeiten, die Natureintheilung aber auf Stämme, welche die Thiere nach Verwandtschaften in Ansehung der Erzeugung eintheilt. Jene verschafft ein Schulsystem für das Gedächtniß; diese ein Natursystem für den Verstand: die erstere hat nur zur Absicht, die Geschöpfe unter Titel, die zweite, sie unter Gesetze zu bringen.

20. In "Bestimmung des Begriffs einer Menschenrace," *Gesammelte Schriften, Band 8,* 102 (italics indicate emphatic spacing):

Denn Thiere, deren Verscheidenheit so groß ist, das zu deren Existenz eben so veil verscheidene Erschaffungen nöthig wären, können wohl zu einer *Nominalgattung* (um sie nach gewissen Ähnlichkeiten zu klassificiren), aber niemals zu einer *Realgattung,* als zu welcher durchaus wenigstens die Möglichkeit der Abstammung von einem einzigen Paar erfodet wird, gehören. . . . Aber auch alsdann würde *zweitens* doch mancher der sonderbare Übereinstimmung de Zeugungskräfte zweier verscheidenen Gattungen die, da sie in Ansehung ihres Ursprungs einander ganz fremd sind, dennoch mit einander fruchtbar vermischt werden können, ganz umsonst und ohne einem anderen Grund, daß es der Natur so gefallen, angenommen werden. Wenn man, um dieses letztere zu beweisen, Thiere anführen, bei denen dieses ungeachtet der Verschedienheit ihres ersten Stammes dennoch, geschehe: so wird ein jeder in solchen Fällen die letztere Voraussetzung leugnen und vielmehr daraus, das eine solche fruchtbare Vermischung statt findet, auf die Einheit des Stammes schleißen, wie aus der Vermischung der Hunde und Füche, u. s. w. Die *unausbleibliche Unartung* beiderseitiger Eigenthümlichkeiten der Eltern ist also der einzig wahre und zugleich hinreichende Probirstein der Verschedienheit der Racen, wozu sie gehören, und ein Beweis der Einheit des Stammes, woraus sie entsprungen sind: nämlich der in diesem Stamm gelegten, sich in der Folge der Zeugungen entwickelden ursprünglichen Keime, ohne welche jene erblichen Mannigsaltigkeiten nicht würden entstanden sein und vornehmlich nicht hätten *nothwendig erblich* können.

21. The page numbers of the second edition are conventionally preceded by the letter *B;* and of the first edition, by the letter *A.*

22. Something I very much doubt. Eco is giving linguistic constraints undue priority here, perhaps in a kind of Sapir-Whorf way. Kant attended to the sciences

because he thought that was where knowledge was gained, empirically. He therefore must think that empirical data can give us new categories.

23. This was, in fact, Aristotle's phrase—"the more and the less"—in his writings, as we have seen.

24. See Polly Winsor's discussions [in Winsor 2004, 2001, 2003] on the distinction between natural and artificial, and essentialist and typological taxonomies. Winsor calls typology the "method of exemplars," which is an apt term. Types applied within species, within genera and within higher taxonomic groups. See also Camardi [2001] for a discussion of the *type* concept. A useful discussion in the early nineteenth century is that of Swainson [1834, chapter 3].

25. Stresemann [1975: 52] notes that Linnaeus is *also* attempting a kind of "natural" system even in his "artificial" system, and he contrasts it to the prior "classical" system—that is, the Aristotelian system of differentiating by general features such as, in the case of birds, land- or water-based lifestyles. As we saw, Bonnet retains a large amount of the classical a priorism of the medievals in this respect.

26. I am indebted to Tom Scharle for the reference and the full quote. Andrew Criddle noted that "Agtsimba" is most likely a corruption of "Agasimba" or "Agasymba," a region vaguely south of Libya (personal communication). This is substantially what Aristotle said in Book VIII of the *Historia animalium*.

THE EARLY NINETEENTH CENTURY:
A PERIOD OF CHANGE

1. Listed as 1851 in the printed version, but this is out of sequence and certainly a typographical error. I am deeply indebted to Mike Dunford for drawing my attention to this comment of Lyell's (and noting the date typo), the cited note of Agassiz's, James Dana's paper, and his conversations with me on the period covered by the "uniformitarian" and "catastrophism" debates in geology. As geology was not, at that time, held to be isolated from any other kind of natural history, Lyell felt, as did Darwin, that the issues raised in the one field (geology) had implications for issues in the other (naturalism). Mike's help has been immense here.

2. Terms like *organized beings*, *organic beings*, *natural beings*, and the like, refer to what we would now call "organisms." Although the term *organism* had been devised in the eighteenth century in French [Cheung 2006], the term was not introduced into English until Owen discussed the kangaroo in 1834, where he said "if the introduction of new powers into an organism necessarily requires a modification in its mode of development" [p. 359], in the context of which it is clear he means a being that has organs, or is organized.

3. My edition is the ninth [1875]. The quoted text is unchanged from the first to the ninth editions.

4. For a more comprehensive overview—but a flawed one in several cases, I believe—of the Benthams' logical enterprise, see McOuat [2003]. The flaw relates

to identifying fixism with essentialism (McOuat agrees in communication) and, to a lesser extent, to not recognizing the much older tradition of the debate over binary privative logic versus multiple species within genera. However, the paper has a much wider agenda, and these are minor problems. Thanks to Charissa Varma for sending me this paper.

5. The terms *intension* and *extension* are apparently medieval, according to Joseph. Mill's *Logic* introduced the terms *connotation* and *denotation* [Joseph 1916: 146–155].

6. Proposed in conversation by Polly Winsor. I have my doubts, though— "essence" is almost always applied by the burgeoning Catholic intelligentsia to knowledge of the nature of God (e.g., by Cardinal John Henry Newman), rather than to physical or material objects. The language was available, but it appears to arise in biology much later. It is also a term in use by continental philosophers influenced by Kant and Hegel—an obvious example is Marx; others include the existentialists such as Kierkegaard (McOuat, personal communication). Nevertheless, I have no doubt that the neo-Thomist revival was an influence on this movement if only because it offered an alternative metaphysics to the Darwinian problem.

7. The French is "L'espèce est une collection ou une suite d'individus caractérisés par un ensemble de traits distinctifs dont la transmission est naturelle, régulière et indéfinie dans l'ordre actuel des choses."

8. Under a section entitled "Lost Species Are Not Varieties of Living Species" [*Règne Animal*, i, 19]:

> Cette recherche suppose la définition de l'espèce qui sert de base à l'usage que l'on fait de ce mot, savoir que l'espèce comprend *les individus qui descendent les uns des autres ou de parens communs, et ceux qui leur ressemblent autant qu'ils se ressemblent entre eux.* [My research assumes the definition of species which serves as the basic use made of the term, understanding that the word species means *the individuals who descend from one another or from common parents and those who resemble them as much as they resemble each other.*]

From http://www.mala.bc.ca/~johnstoi/cuvier.htm, accessed April 20, 2004.

9. Published as Volume I of the *Contributions to the Natural History of the United States*, 1857 to 1862.

10. Disarmingly, Agassiz in the very next chapter refers approvingly to Darwin's work on the coral reefs as a "charming little volume" [p. 154]. Thanks again to Mike Dunford for access to his copy of this work.

11. I have amended it slightly for grammar's sake. The original version was accessed on September 23, 2002, at http://www.athro.com/general/atrans.html.

12. According to Mayr, in Lurie's footnote. It would be interesting to see how those species have fared in the molecular period of systematics.

13. Not *that* modern—the type-token distinction was made by C. S. Peirce [1885; see §§35–37 in Wollheim 1968 for a full discussion of this distinction] only a few decades after Dana wrote. Intriguingly, Peirce's distinction was between *icons*, *indices*, and tokens, and he referred to tokens as *replicas* of symbols [Hookway 1985: 130f.].

14. I am indebted, literally and metaphorically, to Dr. Noelie Alito for purchasing an original copy of Gosse's *Creation (Omphalos)* on my behalf. A scanned copy is available on the Internet from archive.org. The title of physical copies I have seen include both *Omphalos* and the title given here. Possibly the publisher reissued it with another title to increase sales by making the subject matter clearer.

15. Stevens's [1997] article is an excellent source of material and overview for Hooker in particular.

DARWIN AND THE DARWINIANS

1. Possibly it arose from a comment made in 1866 by John Campbell, the Duke of Argyll [1884: 240]: "It will be seen, then, that the principle of Natural Selection has no bearing whatever on the Origin of Species, but only on the preservation and distribution of species when they have arisen. I have already pointed out that Mr. Darwin does not always keep this distinction clearly in view." More likely, though, it is due to the fact that the modern consensus is that species are formed by allopatric isolation, and Darwin held, as we shall see, that they are formed by selection on varieties, now called sympatric speciation. Coyne and Orr, for example, state that he "therefore conflated the problem of change within a lineage with the problem of new lineages" [Coyne and Orr 2004: 11]. I demur: this was Darwin's *hypothesis*, rather than his confusion.

2. All quotations from Darwin's correspondence with Henslow are taken from Barlow [1967].

3. Which is odd, since coral species are notoriously difficult to define or delineate [Carlon and Budd 2002; Pennisi 2002; Soong and Lang 1992; Veron 2001; Vollmer and Palumbi 2002]. At this time he might still have been relying on purely morphological criteria, and this is borne out by the way in which he does describe them.

4. Darwin's views on species do not seem to have changed much between the first edition [Darwin 1859] and this edition. However, his general view on gradual evolution, and the possibility of allopatric speciation, seems to have affected his expression of the nature of species [see Wilkins and Nelson 2008].

5. "There are very variable species and very constant species, and it is obvious that colonies which are founded by a very variable species can hardly ever remain exactly identical with the ancestral species; and that several of them will turn out differently, even granting that the conditions of life be exactly the same, for no colony will contain all the variants of the species in the same proportion, but at most only a few of them, and the result of mingling these must ultimately result in the development of a somewhat different form in each colonial area." [Weismann 1904: 286]

6. *Cladism* is a term that denotes a taxonomic methodology of classifying in terms of shared ancestry or characters that derive from shared ancestry, properly known as phylogenetic systematics, developed in the 1950s. It would be anachronistic to apply to Darwin terms that are only meaningful to describe a later school of thought.

7. A slightly different interpretation is given by Jim Mallet [2008].

8. Two exceptions: Diane Paul [1981] and Stephen Jay Gould [1999]. Gould, however, seems to have wanted to distance himself from the Engels's misinterpretation of Trémaux in order to shield punctuated equilibrium theory from another charge of being unsupportable, and so he says:

> I had long been curious about Trémaux and sought a copy of his book for many years. I finally purchased one a few years ago—and I must say that I have never read a more absurd or more poorly documented thesis. Basically, Trémaux argues that the nature of the soil determines national characteristics and that higher civilizations tend to arise on more complex soils formed in later geological periods. If Marx really believed that such unsupported nonsense could exceed the *Origin of Species* in importance, then he could not have properly understood or appreciated the power of Darwin's facts and ideas. [p. 90]

In fact, my coauthor Gareth Nelson and I think that Marx had the right of it—Trémaux actually is putting forward a mechanism for speciation that explains species (at least, of animals) and human geographic differentiation. By failing to read *sol* as a general term for "habitat" rather than, as Engels did, a term for "geological formations," Gould has unfairly dismissed Trémaux.

9. Wallace is referring, I think, to James Prichard.

THE SPECIES PROBLEM ARISES

1. This series of lectures was collected in 1864 [my edition is Huxley 1895], but the essay was first delivered in 1863.

2. Although Gould [2002: 198] ascribes this to Spencer in 1893, Romanes claimed the honor in his 1895 as an earlier coinage of his own.

3. He defined it [Mayr 1942: 149] thus: "Two forms (or species) are *allopatric*, if they do not occur together, that is if they exclude each other geographically."

4. As we have seen, this anecdote occurs in Darwin's correspondence to Gray in 1857.

5. Jordan [1905b: 157]: "Solche blutsverwandte Individuen bilden eine faunistische Einheit in einem Gebeit," which Jordan follows with the elided comment "zu welcher Einheit wir erfährungsgemäß alle andern Individuen des Gebeits rechnen müssen, welche ihnen gleichen," meaning roughly that we assign all individuals to these units when we empirically classify them as identical. Again, there is an epistemological element here that is overlooked. The subsequent statement Mayr quotes comes after a number of examples of this classification.

6. Thanks to Ian Musgrave for help with the translation: "Das Kriterium des Begriffs Species (= Art) ist daher ein dreifaches, und jeder einzelne Punkt ist der Prüfung zugänglich: Eine Art bat gewisse Körpermerkmale, erzeugt keine den Individuen anderer Arten gleich Nachkommen und verschmilzt nicht mit andern Arten."

7. Page 160: "als ob *nie* ein Zusammenhang zwischen ihnen gewesen, als ob jede Art für sich geschaffen ware." Italics represent emphatic spacing.

THE SYNTHESIS AND SPECIES

1. In order to identify those active in the so-called modern synthesis, which is not looking so modern anymore, we need a term. I trust I can be excused this one.

2. *Polyploidy* is the state of having three or more complete sets of chromosomes, in contrast to the usual state of diploidy in sexual organisms. *Alloploidy* occurs when hybrids are formed through the fertilization of gametes (sex cells) across species. It is usually also polyploidy, in which case it is called allopolyploidy. In plants, fertile individuals often result when chromosomes are duplicated and then separate to form symmetrical diploid chromosome sets. If the hybrid is significantly different from the parental populations, it will not interbreed with them or will interbreed incompletely, so that eventually the novel karyotype (chromosomal type) will breed true with itself but not with either parental type. Speciation can be achieved in one or only a few generations this way [Grant 1975: 431ff.].

3. However, some of the canonical examples are being disputed. Recently, molecular analysis of the *Larus argentatus* (herring gull) complex has indicated that they are, in fact, isolated gene pools [Liebers, Knijff, and Helbig 2004], and the *Parus major* (great tit) complex, while it does interbreed to some extent, is a good set of phylogenetic species [Kvist et al. 2003].

4. In fact, Darwin *explicitly* talked of selection and the struggle for existence as occurring between species, as we have seen.

5. Recent work has (unfortunately for Mayr) shown otherwise [Albertson et al. 1999; Mazeroll and Weiss 1995; Salzburger, Baric, and Sturmbauer 2002; Stauffer, McKaye, and Konings 2002]. Schilthuizen [2000] gives an excellent summary of recent work.

6. Again, recent work has found sufficient examples, mostly in plants and other gamete broadcasters, to establish this as a real process [Aldasoro et al. 1998; Chepurnov et al. 2002; Dowling and Secor 1997; Lee, Mummenhoff, and Bowman 2002; Ramsey, Schemske, and Doyle 1998].

7. According to Chung [2003: 285], Mayr first began to discuss the dimensionality of species in an address in 1946 [Mayr 1946], where he described the Linnaean conception as having no dimensions. Otherwise, his tone is, as Chung remarks, fairly neutral on the difference between the "morphological" species concept and the "biological," "polytypic" species concept at that time. Chung traces Mayr's emerging view of species concepts as differing in their typology and populational nature from 1953 [Mayr, Linsley, and Usinger 1953] through to 1959 [Mayr 1959] and concludes that he discovered the typological aspect of the prior conceptions at around this time [especially in his 1955 work].

8. Apart from flying an "ecological niche" variant in the 1982 *Growth of Biological Thought*, addressed in the section on ecological conceptions.

9. Hull (personal communication) tells the story that, as a graduate student, he delivered the talk on which this paper was based in front of Popper and handed it in at the end of semester. Popper took it on himself to send the talk to the *British Journal for the Philosophy of Science* without Hull's knowledge, and the author had to ask for it to be returned for revision. Many of his conclusions were not

strictly in line with Popper's own ideas, but Popper apparently never read the published work, so Hull never came under Popper's withering attack himself.

10. Polly Winsor believes, after discussing the issue with various of Mayr's students and associates, that Mayr did *not* know Popper's work until his attention was drawn to it by Hull's 1965 paper. Mayr does cite Popper in his later work [Mayr 1997: 59f.], where he explicitly mentions Popper's attitude to words and essentialism, but prior to that, his ideas on typology seem to be his own, as Winsor calls it, his "dragon." According to Winsor, Mayr first used the term *essentialism* as a synonym for *typology* in 1968 [Winsor 2004].

11. Nelson and Platnick's focus on Popper is due to the work of Walter Bock [1974; Nelson, personal communication], although Hull (personal communication) recalls suggesting Popper to one of them. Popper's influence on cladistic taxonomy, including the early cladists, is documented in Rieppel [2003].

12. I am indebted to Neil Thomason for this phrase.

13. Although this may sound harsh, Mayr *has* referred to his "precursors" as "prophetic spirits" [Mayr 1996: 269], noting "how tantalizingly close to a biological species concept some of the earlier authors had come" [Mayr 1982: 271], and claimed that "Buffon understood the gist of it" and the early Darwin also [Mayr 1997: 130], thus claiming authoritative precursors. Hull [1988c: 372–377] and Winsor [2001] discuss the role precursors play in scientific histories. One function for precursors is to give legitimacy to the views of the modern scientist and deflect criticism to dead white males. Similar things happened with Galileo and also with the "rediscoverers" of Mendel.

14. Mayr once told Hull in conversation that not potentially interbreeding was the equivalent of there being isolating mechanisms present (Hull, personal communication).

MODERN DEBATES

1. Thanks to Neil Thomason for bringing this wonderful passage, first published in 1881, to my attention.

2. Hull has noted (in correspondence) that it is ahistorical and whiggist in turn for historians such as myself to apply current standards to the Received Historians of the 1960s; he, Mayr and Cain had what scholarly resources were then available. In large part due to their work, later research has identified the "missing links," and I am certain later work will overturn some of the claims made in this and other modern work, too. It should not be thought that I am criticizing him or Mayr and others for failing to take into account later scholarship.

3. This is the phrase that Sewall Wright used, rather than the later "adaptive landscape," in his initial papers [Wright 1931, 1932] that introduced the idea of genetic drift.

4. Ghiselin suggests the shift from the biospecies emphasis of this definition to the less explicitly interbreeding concept a decade later is due to his falling "increasingly under the spell of the set-theoretical treatment of the Linnaean hierarchy by Gregg . . . , who, although mentioning in passing the possibility that

species are something else, insisted they are classes." However, a set interpretation does not make it, *ipso facto*, a class interpretation.

5. I owe this term to Henry Plotkin (personal communication).

6. There is a rival conception, *holophyly*, which was coined by [Ashlock 1971; see the response by Nelson 1971], to free up the term *monophyly* to mean a looser sense of "clade" in which paraphyly could be tolerated; Hennigian monophyly would thus be holophyly on Ashlock's view. In this writer's opinion, this is a concession to the "evolutionary systematics" of Mayr and collaborators, which seeks to group by similarity as well as genealogy [Mayr and Ashlock 1991].

7. Meier and Willman [2000: 36–37] call the autapomorphic species concept the "Phylogenetic Species Concept *simpliciter*," and Davis [1997] calls the process or synapomorphic species concept the "Autapomorphic Species Concept," in direct contradiction to my usage. I shall use the terms as I here define them, but it should be noted that there are other senses in the literature.

8. The term *process cladism* was introduced in print in Ereshefsky [2000]. It is unclear to me that either pattern cladism or process cladism form monolithic schools, and the differences of opinion on these matters need to investigated. Thanks to David Williams for noting this and catching my transposition of their core ideas.

9. It is not, in my opinion, true that pattern cladism commits its adherents to an antievolutionary view of taxa. Neither is it true that it is an essentialistic view of taxonomy, as some, notably Mayr, have claimed. It is, however, typological.

10. Brent Mishler pointed out to me that what I had been calling a single class of species concepts, under the term "Monophyletic Species Concepts" (now synapomorphic species), was actually pretty diverse, and that some (e.g., Cracraft) did not think species had to be monophyletic. I have attempted to disentangle these views from each other in the text.

11. I am indebted to Dr. Littlejohn for providing me with these references.

12. Viruses can cross over genetic material in a superinfected host [Szathmáry 1992; Boerlijst, Bonhoeffer, and Nowak 1996].

13. The PhyloCode proposes to replace all Linnaean ranks with strictly monophyletic taxa based on the best cladograms [Cantino and de Querioz 2000]. It has been supported by eliminativists like Ereshefsky [2000] and Pleijel and Rouse [2003], but some pattern cladists, such as Norman Platnick and Gareth Nelson (personal communication), oppose it due to its disruption of scientific communication and meaninglessness, as in their view cladograms are only hypotheses and are subject to revision. Others [Benton 2000; Berry 2002; Bryant and Cantino 2002; Carpenter 2003; Forey 2002; Gao and Sun 2003; Keller, Boyd, and Wheeler 2003; Kojima 2003; Nixon 2003] attack the proposal for a range of reasons, ranging from a personal distaste to a rejection of cladism. Most think that named monophyletic higher taxa are not going to be stable as new results come in.

14. A point made at the Melbourne Systematics Forum during 2002. I did not catch the name of the person making the point, but it is important. In the end, unless we know the ways in which DNA is expressed developmentally in all the species being analyzed, DNA is just a richer source of "morphological" data. However, few of these nucleotides are likely, in practice, to be informative (D. Williams, personal communication)

15. Chris Brochu (personal communication).

16. A taphonomist of my acquaintance once noted that taphonomists are people who walk along a deserted beach and, on finding a dead fish or jellyfish, will murmur, "Mmmm, data . . . ."

17. An excellent discussion and a proposed resolution to this problem is Polly [1997]. See also George [1956]; Simpson [1943]; Smith [1994]; Sylvester-Bradley [1956].

18. Texts such as Mayr's and Simpson's [Mayr 1963; Simpson 1961] refer instead to "chronoclines" as the directional change of characters in the paleontological record, akin to geographic clines.

HISTORICAL SUMMARY AND CONCLUSIONS

1. Hull [1967: 312] did quote Aristotle (*De Anima* 415a26) on the generation of like forms. He notes that Aristotle and Theophrastus his student did deal with cases where breeding true did not result in like forms, though. This was not emphasized in Hull's later work.

References

Abulafia, David.
    1988. *Frederick II: A medieval emperor*. London: Allen Lane/Penguin.

Adanson, Michel.
    1763. *Familles des plantes: I. Partie. Contenant une préface istorike sur l'état ancien & actuel de la botanike, & une téorie de cete science*. Paris: Vincent.

Agassiz, Louis.
    1842. New views regarding the distribution of fossils in formations. *Edinburgh New Philosophical Journal* 32(63): 97–98.
    1859. *An essay on classification*. London: Longman, Brown, Green, Longmans and Roberts and Trubner.
    1863. *Methods of study in natural history*. Boston: Ticknor and Fields.
    1869. *De l'Espece et de la classification en zoologie*. Paris: Balliere.

Aguilar, Javier Fuertes, Josep Antoni Rosselló, and Gonzalo Nieto Feliner.
    1999. Molecular evidence for the compilospecies model of reticulate evolution in *Armeria* (Plumbaginaceae). *Systematic Biology* 48(4): 735–754.

Albertson, R. C., J. A. Markert, P. D. Danley, and T. D. Kocher.
    1999. Phylogeny of a rapidly evolving clade: The cichlid fishes of Lake Malawi, East Africa. *Proceedings of the National Academy of Sciences of the USA* 96: 5107–5110.

Albertus Magnus.
    1987. *Man and the beasts: De animalibus (Books 22–26)*. Translated by J. J. Scanlan. Binghamton, NY: Medieval & Renaissance Texts & Studies.

Aldasoro, J. J., C. Aedo, C. Navarro, and F. M. Garmendia.
1998. The genus *Sorbus* (Maloideae, Rosaceae) in Europe and in North Africa: Morphological analysis and systematics. *Systematic Botany* 23(2): 189–212.

Allen, Garland E.
1980. The evolutionary synthesis: Morgan and natural selection revisited. In *The evolutionary synthesis*, edited by E. Mayr and W. B. Provine. New York: Columbia University Press: 356–382.

Amundson, Ron.
1996. Historical development of the concept of adaptation. In *Adaptation*, edited by M. R. Rose and G. V. Lauder. San Diego: Academic Press: 11–53.
1998. Typology reconsidered: Two doctrines on the history of evolutionary biology. *Biology and Philosophy* 13(2): 153–177.
2005. *The changing rule of the embryo in evolutionary biology: Structure and synthesis, Cambridge studies in philosophy and biology*. New York: Cambridge University Press.

Arber, Agnes.
1938. *Herbals: Their origin and evolution: A chapter in the history of botany 1470–1670*. 2nd ed. Cambridge: Cambridge University Press.

Argyll, George J. D. Campbell, the Duke of.
1884. *The reign of law*. 18th ed. London: Alexander Strahan. Original edition, 1866.

Aristotle.
1998. *The Metaphysics*. Translated by H. Lawson-Tancred. London: Penguin.

Ashlock, Peter D.
1971. Monophyly and associated terms. *Systematic Zoology* 20(1): 63–69.

Atran, Scott.
1985. The early history of the species concept: An anthropological reading. In *Histoire du concept d'espece dans les sciences de la vie*. Paris: Fondation Singer-Polignac: 1–36.
1990. *The cognitive foundations of natural history*. New York: Cambridge University Press.
1995. Causal constraints on categories and categorical constraints on biological reasoning across cultures. In *Causal cognition: A multidisciplinary debate*, edited by D. Sperber, D. Premack, and A. J. Premack. Oxford: Clarendon Press.
1998. Folk biology and the anthropology of science: Cognitive universals and the cultural particulars. *Behavioral and Brain Sciences* 21(4): 547–609.
1999. The universal primacy of generic species in folkbiological taxonomy: Implications for human biological, cultural and scientific evolution. In *Species: New interdisciplinary essays*, edited by R. A. Wilson. Cambridge, MA: Bradford/MIT Press: 231–261.

Augustine, Saint, Bishop of Hippo.
1962. *The City of God*. Translated by J. Healey and R. V. G. Tasker. London: Dent.
1982. *The literal meaning of Genesis*. Translated by J. H. Taylor, *Ancient Christian writers; no. 41–42*. New York: Newman Press.

Avise, J. C., and R. M. Ball Jr.
1990. Principles of genealogical concordance in species concepts and biological taxonomy. In *Oxford Surveys in Evolutionary Biology*, edited by D. Futuyma and J. Atonovics. Oxford: Oxford University Press: 45–67.

Avise, J. C., and K. Wollenberg.
1997. Phylogenetics and the origin of species. *Proceedings of the National Academy of Sciences of the USA* 94(15): 7748–7755.

Ayala, Francisco José.
1982. Gradualism versus punctuationism in speciation: Reproductive isolation, morphology, genetics. In *Mechanisms of speciation: Proceedings from the international meeting on mechanisms of speciation, sponsored by the Academia Nazionale dei Lincei, May 4–8, 1981, Rome, Italy*, edited by C. Barigozzi. New York: Liss.

Bacon, Francis.
1960. *The new Organon and related writings*. Translated by F. H. Anderson. Indianapolis: Bobbs-Merrill. Original edition, 1620.

Baldauf, S. L., J. D. Palmer, and W. F. Doolittle.
1996. The root of the universal tree and the origin of eukaryotes based on elongation factor phylogeny. *Proceedings of the National Academy of Sciences of the USA* 93: 7749–7754.

Baldwin, James Mark, ed.
1901. *Dictionary of philosophy and psychology including many of the principal conceptions of ethics, logic, aesthetics, philosophy of religion, mental pathology, anthropology, biology, neurology, physiology, economics, political and social philosophy, philology, physical science, and education and giving a terminology in English, French, German, and Italian*. 3 vols. New York: Macmillan.

Balme, D. M.
1987a. Aristotle's biology was not essentialist. In *Philosophical issues in Aristotle's biology*, edited by A. Gotthelf and J. G. Lennox. Cambridge: Cambridge University Press: 291–312.
1987b. The place of biology in Aristotle's philosophy. In *Philosophical issues in Aristotle's biology*, edited by A. Gotthelf and J. G. Lennox. Cambridge: Cambridge University Press: 9–20.

Barlow, Nora, ed.
1967. *Darwin and Henslow: The growth of an idea: Letters, 1831–1860*. London: Murray [for] Bentham-Moxon Trust.

Barnes, J., ed.
1984. *The complete works of Aristotle*. 2 vols. Princeton, NJ: Princeton University Press.

Barrington, D. S., C. H. Haufler, and C. R. Werth.
1989. Hybridization, reticulation, and species concepts in the ferns. *American Fern Journal* 79(2): 55–64.

Basalla, George, William Coleman, and Robert H. Kargon, eds.
1970. *Victorian science: A self-portrait from the presidential addresses of the British Association for the Advancement of Science*. New York: Anchor/Doubleday.

Bateson, William.
    1894. *Material for the study of variation treated with especial regard to discontinuity in the origin of species.* London: Macmillan.

Beatty, J.
    1985. Speaking of species: Darwin's strategy. In *The Darwinian heritage*, edited by D. Kohn. Princeton, NJ: Princeton University Press.

Beckner, M.
    1959. *The biological way of thought.* New York: Columbia University Press.

Beltran, M., C. D. Jiggins, V. Bull, M. Linares, J. Mallet, W. O. McMillan, and E. Bermingham.
    2002. Phylogenetic discordance at the species boundary: Comparative gene genealogies among rapidly radiating heliconius butterflies. *Molecular and Biological Evolution* 19(12): 2176–90.

Bentham, George.
    1827. *An outline of a new system of logic: With a critical examination of Dr. Whately's "Elements of Logic."* London: Hunt and Clark.

Benton, M. J.
    2000. Stems, nodes, crown clades, and rank-free lists: Is Linnaeus dead? *Biological Reviews* 75(4): 633–648.

Berry, P. E.
    2002. Biological inventories and the PhyloCode. *Taxon* 51(1): 27–29.

Bigelow, R. S.
    1965. Hybrid zones and reproductive isolation. *Evolution* 19(4): 449–458.

Blackmore, John.
    1979. On the inverted use of the terms 'realism' and 'idealism' among scientists and historians of science. *British Journal for the History of Science* 30: 125–134.

Blackwelder, Richard E.
    1967. *Taxonomy: a text and reference book.* New York: Wiley.

Bock, Walter J.
    1974. Philosophical foundations of classical evolutionary classification. *Systematic Zoology* 22: 375–392.

Boerlijst, M. C., S. Bonhoeffer, and M. A. Nowak.
    1996. Viral quasispecies and recombination. *Proceedings of the Royal Society of London, Series B* 263: 1577–1584.

Boodin, John Elof.
    1943. The discovery of form. *Journal of the History of Ideas* 4(2): 177–192.

Bowler, Peter J.
    1983. *The eclipse of Darwinism: Anti-Darwinian evolution theories in the decades around 1900.* Baltimore, MD: Johns Hopkins University Press.
    1989a. *Evolution: the history of an idea.* Rev. ed. Berkeley: University of California Press. Original edition, 1984.
    1989b. *The Mendelian revolution: The emergence of hereditarian concepts in modern science and society.* Baltimore, MD: Johns Hopkins University Press.

Brande, W. T., and Joseph Cauvin, eds.
    1853. *A dictionary of science, literature, and art: Comprising the history, description, and scientific principles of every branch of human knowledge, with the derivation and definition of all the terms in general use.* 2nd ed. London: Longman.

Bridle, J. R., and M. G. Ritchie.
    2001. Assortative mating and the genic view of speciation. *Journal of Evolutionary Biology* 14(6): 878–879.

Broberg, Gunnar.
    1983. *Homo sapiens*: Linnaeus's classification of man. In *Linnaeus, the man and his work*, edited by T. Frängsmyr. Berkeley: University of California Press: 156–194.

Brower, Andrew V. Z.
    2002. Cladistics, populations and species in geographical space: The case of *Heliconius* butterflies. In *Molecular systematics and evolution: Theory and practice*, edited by R. DeSalle, G. Girbet, and W. Wheeler. Basel: Birkhäuser: 5–15.

Brumbaugh, Robert S.
    1981. *The philosophers of Greece.* Albany: State University of New York Press.

Bryant, H. N., and P. D. Cantino.
    2002. A review of criticisms of phylogenetic nomenclature: Is taxonomic freedom the fundamental issue? *Biological Reviews* 77(1): 39–55.

Bulmer, Ralph.
    1967. Why is the cassowary not a bird? A problem among the Karam of the New Guinea highlands. *Journal of the Royal Anthropological Institute* 2(1): 5–25.

Burg, T. M.
    1999. Isolation and characterization of microsatellites in albatrosses. *Molecular Ecology* 8(2): 338–41.

Burg, T. M., and J. P. Croxall.
    2004. Global population structure and taxonomy of the wandering albatross species complex. *Molecular Ecology* 13(8): 2345–55.

Burkhardt, Frederick, ed.
    1996. *Charles Darwin's letters: A selection, 1825–1859.* Cambridge: Cambridge University Press.

Butlin, R., and M. G. Ritchie.
    2001. Evolutionary biology: Searching for speciation genes. *Nature* 412(6842): 31, 33.

Butterfield, Herbert.
    1931. *The Whig interpretation of history.* London: Bell.

Cain, Arthur J.
    1954. *Animal species and their evolution.* London: Hutchinson University Library.

1958. Logic and memory in Linnaeus's system of taxonomy. *Proceedings of the Linnean Society of London* 169: 144–163.

ed. 1959a. *Function and taxonomic importance: a symposium.* London: Systematics Association.

1959b. The post-Linnaean development of taxonomy. *Proceedings of the Linnean Society of London* 170: 234–244.

1959c. Taxonomic concepts. *Ibis* 101: 302–318.

1993. Linnaeus's Ordines naturales. *Archives of Natural History* 20: 405–415.

1994. Numerus, figura, proportio, situs: Linnaeus's definitory attributes. *Archives of Natural History* 21: 17–36.

1995. Linnaeus's natural and artificial arrangements of plants. *Botanical Journal of the Linnean Society* 117(2): 73.

1997. John Locke on species. *Archives of Natural History* 24(3): 337–360.

1999a. John Ray on the species. *Archives of Natural History* 26(2): 223–238.

1999b. Thomas Sydenham, John Ray, and some contemporaries on species. *Archives of Natural History* 24(1): 55–83.

Camardi, Giovanni.

2001. Richard Owen, morphology and evolution. *Journal of the History of Biology* 34(3): 481.

Campbell, Donald T.

1965. Variation and selective retention in socio-cultural evolution. In *Social change in developing areas: A reinterpretation of evolutionary theory,* edited by H. R. Barringer, G. I. Blanksten, and R. W. Mack. Cambridge, MA: Schenkman.

Candolle, Augustine-Pyramus de.

1819. *Théorie élementaire de la botanique, ou exposition des principes de la classification naturelle et de l'art de décrire et d'étudier les végétaux.* 2nd ed. Paris: N.p.

Cantino, Philip D., and Kevin de Querioz.

2000. *Phylocode: A phylogenetic code of biological nomenclature.* Available from http://www.ohiou.edu/phylocode/.

Carlon, D. B., and A. F. Budd.

2002. Incipient speciation across a depth gradient in a scleractinian coral? *Evolution: International Journal of Organic Evolution* 56(11): 2227–42.

Carpenter, J. M.

2003. Critique of pure folly. *Botanical Review* 69(1): 79–92.

Carroll, Lewis.

1962. *Alice's Adventures in Wonderland, and, Through the looking glass.* Harmondsworth: Penguin. Original edition, 1865/1871.

Carson, Hampton L.

1957. The species as a field for gene recombination. In *The species problem: A symposium presented at the Atlanta meeting of the American Association for the Advancement of Science, December 28–29, 1955,* Publication No 50, edited by E. Mayr. Washington DC: American Association for the Advancement of Science: 23–38.

1971. Speciation and the founder principle. *Stadler Genetics Symposium* 3: 51–70.

1975. The genetics of speciation at the diploid level. *American Naturalist* 109: 83–92.

Carson, Hampton L., D. E. Hardy, H. T. Spieth, and W. S. Stone.

1970. The evolutionary biology of the Hawaiian *Drosophilidae*. In *Essays in honor of Theodosius Dobzhansky*. New York: Appleton-Century-Crofts: 437–543.

Cassirer, Ernst, Paul Oskar Kristeller, and John Herman Randall, eds.

1948. *The Renaissance philosophy of man: Selections in translation*. Chicago: University of Chicago Press.

Chambers, Robert.

1844. *Vestiges of the natural history of creation*. London: Churchill.

Charles, David.

2002. *Aristotle on Meaning and Essence*. Oxford: Oxford University Press.

Chepurnov, Victor A., David G. Mann, Wim Vyverman, Koen Sabbe, and Daniel B. Danielidis.

2002. Sexual reproduction, mating system, and protoplast dynamics of Seminavis (Bacillariophyceae). *Journal of Phycology* 38(5): 1004–1019.

Cheung, Tobias.

2006. From the organism of a body to the body of an organism: Occurrence and meaning of the word from the seventeenth to the nineteenth centuries. *British Journal for the History of Science* 39(03): 319–339.

Chung, Carl.

2003. On the origin of the typological/population distinction in Ernst Mayr's changing views of species, 1942–1959. *Studies in History and Philosophy of Biological and Biomedical Sciences* 34: 277–296.

Clark, Brett, John Foster, and Richard York.

2007. The critique of intelligent design: Epicurus, Marx, Darwin, and Freud and the materialist defense of science. *Theory and Society* 36(6): 515–546.

Clarke, Richard F.

1895. *Logic*. 3rd ed, *Manuals of Catholic Philosophy*. London: Longmans, Green.

Clauss, Sidonie.

1982. John Wilkins' *Essay toward a Real Character*: Its place in seventeenth-century episteme. *Journal of the History of Ideas* 43(4): 531–553.

Coggon, Jennifer.

2002. Quinarianism after Darwin's *Origin*: The circular system of William Hincks. *Journal of the History of Biology* 35(1): 5.

Cohan, Frederick M.

2001. Bacterial species and speciation. *Systematic Biology* 50(4): 513–524.

2002. What are bacterial species? *Annual Review of Microbiology* 56: 457–487.

2006. Towards a conceptual and operational union of bacterial systematics, ecology, and evolution. *Philosophical Transactions of the Royal Society B: Biological Sciences* 361(1475): 1985–1996.

Collingwood, R. G.
   1946. *The idea of history.* Oxford: Oxford University Press.

Cope, Edward Drinker.
   1868. On the origin of genera. *Proceedings of the Academy of Natural Sciences of Philadelphia* 20: 242–300.

Coyne, Jerry A.
   1994. Ernst Mayr and the origin of the species. *Evolution* 48(1): 19–30.

Coyne, Jerry A., and H. Allen Orr.
   2004. *Speciation.* Sunderland, MA: Sinauer.

Cracraft, Joel.
   1983. Species concepts and speciation analysis. In *Current Ornithology*, edited by R. F. Johnston. New York: Plenum Press: 159–187.
   1997. Species concepts in systematics and conservation biology: An ornithological viewpoint. In *Species: The units of biodiversity*, edited by M. F. Claridge, H. A. Dawah, and M. R. Wilson. London: Chapman and Hall: 325–339.
   2000. Species concepts in theoretical and applied biology: A systematic debate with consequences. In *Species concepts and phylogenetic theory: A debate*, edited by Q. D. Wheeler and R. Meier. New York: Columbia University Press: 3–14.

Croizat, Leon.
   1945. History and nomenclature of the higher units of classification. *Bulletin of the Torrey Botanical Club* 72(1): 52–75.

Cronquist, A.
   1978. Once again, what is a species? In *BioSystematics in Agriculture*, edited by L. Knutson. Montclair, NJ: Alleheld Osmun: 3–20.

Curley, Michael J., ed.
   1979. *Physiologus.* Austin: University of Texas Press.

Cuvier, Georges.
   1812. *Discours sur les révolutions du globe (Discourse on the revolutionary upheavals on the surface of the earth).* Translated by I. Johnston. Paris: N.p.
   1835. Éloge de M. de Lamarck. *Mémoires de l'Académie Royale des Sciences de l'Institut de France*, 2nd series, 13: i–xxxi.

Dana, James D.
   1857. Thoughts on species. *American Journal of Science and Arts* 24(72): 305–316.

Darlington, Cyril Dean.
   1940. Taxonomic species and genetic systems. In *The new systematics*, edited by J. Huxley. London: Oxford University Press: 137–160.

Darwin, Charles.
   1839. *Journal of researches into the geology and natural history of the various countries visited by H.M.S. Beagle under the command of Captain Fitzroy, R.N. from 1832 to 1836.* London: Colburn.
   1842. *The structure and distribution of coral reefs: Being the first part of the Geology of the voyage of the Beagle, under the command of Capt. Fitzroy, R.N., during the years 1832 to 1836.* London: Murray.

1845. *Journal of researches into the natural history and geology of the countries visited during the voyage of H.M.S. Beagle round the world: Under the command of Capt. Fitz Roy*. 2nd ed., corr. with additions ed. London: Murray.

1851. *A monograph on the sub-class Cirripedia: With figures of all the species*. London: Printed for the Ray Society.

1859. *On the origin of species by means of natural selection, or The preservation of favoured races in the struggle for life*. London: Murray.

1871. *The descent of man and selection in relation to sex*. London: Murray.

1872. *The origin of species by means of natural selection, or, The preservation of favoured races in the struggle for life*. 6th ed. London: Murray.

1873. Origin of certain instincts. *Nature* vii: 417.

1875. *The variation of animals and plants under domestication*. 2nd rev. ed. London: Murray. Original edition, 1868.

1972. *More letters of Charles Darwin: A record of his work in a series of hitherto unpublished letters*. Edited by A. Francis Darwin and A. C. Seward. New York: Johnson Reprint. Original edition, 1903.

1975. *Charles Darwin's natural selection: Being the second part of his big species book written from 1856 to 1858*. Edited by R. C. Stauffer. London: Cambridge University Press.

1998. *The variation of animals and plants under domestication*. 2nd ed. 2 vols. Baltimore, MD: Johns Hopkins University Press. Original edition, 1868.

Darwin, Francis, ed.

1888. *The life and letters of Charles Darwin: Including an autobiographical chapter*. 3 vols. London: Murray.

Davis, Jerrold I.

1997. Evolution, evidence, and the role of species concepts in phylogenetics. *Systematic Biology* 22(2): 373–403.

de Queiroz, Kevin.

1998. The general lineage concept of species, species criteria, and the process of speciation. In *Endless forms: Species and speciation*, edited by D. J. Howard and S. H. Berlocher. New York: Oxford University Press: 57–75.

1999. The general lineage concept of species and the defining properties of the species category. In *Species, New interdisciplinary essays*, edited by R. A. Wilson. Cambridge, MA: Bradford/MIT Press: 49–88.

2007. Species concepts and species delimitation. *Systematic Biology* 56(6): 879–886.

de Queiroz, Kevin, and Michael J. Donoghue.

1988. Phylogenetic systematics and the species problem. *Cladistics* 4: 317–338.

1990. Phylogenetic systematics and species revisited. *Cladistics* 6: 83–90.

de Vries, Hugo.

1901. *Die Mutationstheorie: Versuche und Beobachtungen über die Entstehung von Arten im Pflanzenreich*. 2 vols. Leipzig: Veit.

1912. *Species and varieties: Their origin by mutation: Lectures delivered at the University of California*. 3rd ed. Chicago: Open Court. Original edition, 1904.

Depew, David J., and Bruce H. Weber.

    1995. *Darwinism evolving: Systems dynamics and the genealogy of natural selection*. Cambridge, MA: MIT Press.

Desmond, Adrian J.

    1984. *Archetypes and ancestors: Palaeontology in Victorian London, 1850–1875*. Chicago: University of Chicago Press.

Desmond, Adrian, and James Moore.

    1991. *Darwin*. Harmondsworth, UK: Penguin.

Dewey, John.

    1997. *"The Influence of Darwin on Philosophy" and other essays: Great books in philosophy*. Amherst, NY: Prometheus Books.

Dioscorides, Pedanius, of Anazarbos.

    1959. *The Greek herbal of Dioscorides: Illustrated by a Byzantine, A.D. 512*. Translated by J. Goodyer and R. W. T. Gunther. New York: Hafner.

Dobzhansky, Theodosius.

    1935. A critique of the species concept in biology. *Philosophy of Science* 2: 344–355.

    1937a. *Genetics and the origin of species*. New York: Columbia University Press.

    1937b. What is a species? *Scientia* 61: 280.

    1941. *Genetics and the origin of species*. 2nd ed. New York: Columbia University Press.

    1951. *Genetics and the origin of species*. 3rd rev. ed. New York: Columbia University Press.

    1962. *Mankind evolving: The evolution of the human species*. New Haven, CT: Yale University Press.

    1980. Morgan and his school in the 1930s. In *The evolutionary synthesis*, edited by E. Mayr and W. B. Provine. New York: Columbia University Press: 445–452.

Dowling, Thomas E., and Carol L. Secor.

    1997. The role of hybridization and introgression in the diversification of animals. *Annual Review of Ecology and Systematics* 28: 593–619.

Dres, M., and J. Mallet.

    2002. Host races in plant-feeding insects and their importance in sympatric speciation. *Philosophical Transactions of the Royal Society of London B, Biological Sciences* 357(1420): 471–92.

Eco, Umberto.

    1999. *Kant and the Platypus: Essays on language and cognition*. London: Vintage/Random House.

Eddy, John H., Jr.

    1994. Buffon's *Histoire naturelle*: History? A critique of recent interpretations. *Isis* 85: 644–61.

Eigen, Manfred.

    1993a. The origin of genetic information: viruses as models. *Gene* 135(1–2): 37–47.

    1993b. Viral quasispecies. *Scientific American* 1993 (July): 32–39.

Eldredge, Niles.
  1985. *Time frames: The evolution of punctuated equilibria.* Princeton, NJ: Princeton University Press.
  1989. *Macroevolutionary dynamics: Species, niches, and adaptive peaks.* New York: McGraw-Hill.
  1993. What, if anything, is a species? In *Species, species concepts, and primate evolution,* edited by W. H. Kimbel and L. B. Martin. New York: Plenum Press: 3–20.

Eldredge, Niles, and Joel Cracraft.
  1980. *Phylogenetic patterns and the evolutionary process: Method and theory in comparative biology.* New York: Columbia University Press.

Eldredge, Niles, and Stephen J. Gould.
  1972. Punctuated equilibria: An alternative to phyletic gradualism. In *Models in paleobiology,* edited by T. J. M. Schopf. San Francisco: Freeman Cooper: 82–115.

Eldredge, Niles, Stephen Jay Gould, Jerry A. Coyne, and Brian Charlesworth.
  1997. On punctuated equilibria. *Science* 276 (April 18): 337c–341c.

Ellen, Roy F.
  1993. *The cultural relations of classification: An analysis of Nuaulu animal categories from central Seram.* Cambridge: Cambridge University Press.

England, Richard.
  1997. Natural selection before the *Origin:* Public reactions of some naturalists to the Darwin-Wallace papers (Thomas Boyd, Arthur Hussey, and Henry Baker Tristram). *Journal of the History of Biology* 30(2): 267–290.

Ereshefsky, Marc.
  1999. Species and the Linnean hierarchy. In *Species: New interdisciplinary essays,* edited by R. A. Wilson. Cambridge, MA: Bradford/MIT Press: 285–305.
  2000. *The poverty of Linnaean hierarchy: A philosophical study of biological taxonomy.* Cambridge: Cambridge University Press.

Estes, W. K.
  1994. *Classification and cognition.* New York: Oxford: Oxford University Press.

Farber, Paul Lawrence.
  1971. Buffon's concept of species. PhD diss., Indiana University, Bloomington.
  1976. The type-concept in zoology during the first half of the nineteenth century. *Journal of the History of Biology* 9(1): 93–119.

Farley, John.
  1977. *The spontaneous generation controversy from Descartes to Oparin.* Baltimore, MD: Johns Hopkins University Press.

Fisher, Ronald Aylmer.
  1930. *The genetical theory of natural selection.* Oxford: Clarendon Press.

Flaubert, Gustave.
  1976. *Bouvard and Pécuchet, with the Dictionary of Received Ideas.* Translated by A. J. Krailsheimer. Harmondsworth, UK: Penguin. Original edition, 1881.

Forey, P. L.
   2002. PhyloCode—Pain, no gain. *Taxon* 51(1): 43–54.

Forsdyke, Donald R.
   2001. *The origin of species revisited: A Victorian who anticipated modern developments in Darwin's theory*. Kingston, Ontario: McGill-Queen's University Press.

Foucault, Michel.
   1970. *The order of things: An archaeology of the human sciences*. London: Routledge Classics.

Frängsmyr, Tore, ed.
   1983. *Linnaeus, the man and his work*. Berkeley: University of California Press.

Freedman, Joseph S.
   1993. The diffusion of the writings of Petrus Ramus in Central Europe, c. 1570–c. 1630. *Renaissance Quarterly* 46(1): 98–152.

Futuyma, Douglas J.
   1983. *Science on trial: The case for evolution*. New York: Pantheon.

Gao, K. Q., and Y. L. Sun.
   2003. Is the PhyloCode better than Linnaean system? New development and debate on biological nomenclatural issues. *Chinese Science Bulletin* 48(3): 308–312.

Gasking, Elizabeth B.
   1967. *Investigations into generation 1651–1828: History of scientific ideas*. London: Hutchinson.

Gayon, Jean.
   1996. The individuality of the species: A Darwinian theory? From Buffon to Ghiselin, and back to Darwin. *Biology and Philosophy* 11: 215–244.

George, T. N.
   1956. Biospecies, chronospecies and morphospecies. In *The species concept in paleontology*, edited by P. C. Sylvester-Bradley. London: Systematics Association: 123–137.

Ghiselin, Michael T.
   1966. On psychologism in the logic of taxonomic controversies. *Systematic Zoology* 15: 207–215.
   1969. *The triumph of the Darwinian method*. Berkeley: University of California Press.
   1974a. *The economy of nature and the evolution of sex*. Berkeley: University of California Press.
   1974b. A radical solution to the species problem. *Systematic Zoology* 23: 536–544.
   1984. *The triumph of the Darwinian method, with a new preface*. Rev. ed. Chicago: University of Chicago Press. Original edition, 1969.
   1997. *Metaphysics and the origin of species*. Albany: State University of New York Press.

Gil-White, Francisco.
2001. Are ethnic groups biological "species" to the human brain? Essentialism in our cognition of some social categories. *Current Anthropology: A World Journal of the Human Sciences* 42(4): 515–554.

Gillispie, Charles Coulston.
1959. Lamarck and Darwin in the history of science. In *Forerunners of Darwin 1749–1859*, edited by B. Glass, O. Temkin and W. L. Straus. Baltimore, MD: Johns Hopkins University Press: 265–291.

Gilmour, J. S. L.
1958. The Species: Yesterday and To-Morrow. *Nature* 181 (4606): 379–380.

Glass, Bentley.
1959a. The germination of the idea of biological species. In *Forerunners of Darwin, 1745–1859*, edited by B. Glass, O. Temkin and W. L. Straus Jr. Baltimore, MD: Johns Hopkins University Press: 30–48.
1959b. Heredity and variation in the eighteenth century concept of the species. In *Forerunners of Darwin, 1745–1859*, edited by B. Glass, O. Temkin, and W. L. Straus Jr. Baltimore, MD: Johns Hopkins University Press: 144–172.

Godron, Dominique-Alexandre.
1854. *Florula Juvenalis ou énumération des plantes étrangères qui croissent naturellement au Port Juvénal, près de Montpellier, précédée de considérations sur les migrations des végétaux.* 2nd ed. Nancy: Grimblot et Veuve Raybois.

Goerke, Heinz.
1973. *Linnaeus.* New York: Scribner.

Goldschmidt, Richard B.
1940. *The material basis of evolution.* Seattle: University of Washington Press.

Gosse, Edmund.
1970. *Father and son: A study of two temperaments.* London: Heinemann. Original edition, Heinemann, 1907.

Gosse, Philip Henry.
1857. *Creation (Omphalos): an attempt to untie the geological knot.* London: Van Voorst.

Gould, Stephen Jay.
1982. Darwinism and the expansion of evolutionary theory. *Science* 216(4544): 380–387.
1993. *Eight little piggies: Reflections in natural history.* New York: Norton.
1994. Tempo and mode in the macroevolutionary reconstruction of Darwinism. *Proceedings of the National Academy of Sciences of the USA* 91(15): 6764–6771.
1999. A Darwinian gentleman at Marx's funeral. *Natural History* 108(7): 32–41.
2002. *The structure of evolutionary theory.* Cambridge, MA: Belknap Press of Harvard University Press.

Gould, Stephen Jay, and Niles Eldredge.
1977. Punctuated equilibria: The tempo and mode of evolution reconsidered. *Paleobiology* 3: 115–151.

Grant, Verne.

    1975. *Genetics of flowering plants*. New York: Columbia University Press.

Gray, Samuel Frederick.

    1821. *A natural arrangement of British plants: According to their relation to each other, as pointed out by Jussieu, De Candolle, Brown, & c. including those cultivated for use: with an introduction to botany in which the terms newly introduced are explained*. London: Baldwin, Cradock, and Joy.

Green-Pedersen, Niels Jørgen.

    1984. *The tradition of the Topics in the Middle Ages*. Munich: Philosophia.

Greene, John C.

    1959. *The death of Adam: Evolution and its impact on Western thought*. Ames: Iowa State University Press.

    1963. *Darwin and the modern world view: The Rockwell Lectures, Rice University*. New York: New American Library. Original edition, 1961.

Grew, Nehemiah.

    1682. *The anatomy of plants with an idea of a philosophical history of plants, and several other lectures, read before the Royal Society, Early English Books, 1641–1700/456:17*. London: Rawlins.

Griffiths, Paul E.

    1997. *What emotions really are: The problem of psychological categories*. Chicago: University of Chicago Press.

Hacking, Ian.

    1983. *Representing and intervening: Introductory topics in the philosophy of natural science*. Cambridge: Cambridge University Press.

Haeckel, Ernst Heinrich Philipp August.

    1896. *The evolution of man: A popular exposition of the principal points of human ontogeny and phylogeny*. 3rd ed. New York: Appleton.

Hagberg, Knut.

    1952. *Carl Linnaeus*. Translated by A. Blair. London: Cape.

Hájek, Alan.

    2003. Interpretations of Probability. In *The Stanford Encyclopedia of Philosophy*, edited by E. N. Zalta. http://plato.stanford.edu/archives/sum2003/entries/probability-interpret/.

Haldane, J. B. S.

    1956. Can a species concept be justified? In *The species concept in palaeontology: A symposium*, edited by P. C. Sylvester-Bradley. London: Systematics Association: 95–96.

Hamilton, William, Henry Longueville Mansel, and John Veitch.

    1874. *Lectures on metaphysics and logic*. Edinburgh: Blackwood.

Harlan, J. R., and J. M. J. De Wet.

    1963. The compilospecies concept. *Evolution* 17: 497–501.

Haufler, Christopher.

    1996. Species concepts and speciation in pteridophytes. In *Pteridology in perspective*, edited by J. M. Camus, M. Gibby, and R. J. Johns. Kew, UK: Royal Botanic Gardens.

Hennig, Willi.

1950. *Grundzeuge einer Theorie der Phylogenetischen Systematik*. Berlin: Aufbau.

1966. *Phylogenetic systematics*. Translated by D. D. Davis and R. Zangerl. Urbana: University of Illinois Press.

Hey, J., and J. Wakeley.

1997. A coalescent estimator of the population recombination rate. *Genetics* 145(3): 833–846.

Hey, Jody.

2001a. *Genes, concepts and species: The evolutionary and cognitive causes of the species problem*. New York: Oxford University Press.

2001b. The mind of the species problem. *Trends in Ecology & Evolution* 16(7): 326–329.

Hookway, Christopher.

1985. *Peirce: The arguments of the philosophers*. London: Routledge & Kegan Paul.

Hopkins, Jasper.

1981. *Nicholas of Cusa on learned ignorance: A translation and an appraisal of* De docta ignorantia. Minneapolis: Benning.

Hull, David L.

1965. The effect of essentialism on taxonomy: Two thousand years of stasis. *British Journal for the Philosophy of Science* 15: 314–326, 16: 1–18.

1967. The metaphysics of evolution. *British Journal for the History of Science* 3(12): 309–337.

ed. 1973a. *Darwin and his critics: The reception of Darwin's theory of evolution by the scientific community*. Cambridge, MA: Harvard University Press.

1973b. A populational approach to scientific change. *Science* 182: 1121–1124.

1976. Are species really individuals? *Systematic Zoology* 25: 174–191.

1980. Individuality and selection. *Annual Review of Ecology and Systematics* 11: 311–332.

1981. Units of evolution: A metaphysical essay. In *The philosophy of evolution*, edited by U. L. Jensen and R. Harré. Brighton: Harvester Press: 23–44.

1983a. Conceptual evolution and the eye of the octopus. Paper read at Proceedings of the Seventh International Congress of Logic, Methodology, and Philosophy of Science, Salzburg, Austria.

1983b. Darwin and the nature of science. In *Evolution from molecules to men*, edited by D. S. Bendall. Cambridge: Cambridge University Press.

1984a. Cladistic theory: Hypotheses that blur and grow. In *Cladistic perspectives on the reconstruction of evolutionary history*, edited by T. Duncan and T. Stuessy. New York: Columbia University Press: 5–23.

1984b. Darwinism as a historical entity. In *The Darwinian heritage*, edited by D. Kohn. Wellington, New Zealand: Nova Pacifica.

1984c. Historical entities and historical narratives. In *Minds, machines, and evolution*, edited by C. Hookway. Cambridge: Cambridge University Press.

1988a. A mechanism and its metaphysics: An evolutionary account of the social and conceptual development of science. *Biology and Philosophy* 3: 123–155.

1988b. A period of development: A response. *Biology and Philosophy* 3: 241–261.

1988c. *Science as a process: An evolutionary account of the social and conceptual development of science.* Chicago: University of Chicago Press.

1989a. A function for actual examples in philosophy of science. In *What the philosophy of biology is: Essays dedicated to David Hull,* edited by M. Ruse. Dordrecht: Kluwer: 309–321.

1989b. *The metaphysics of evolution.* Albany: State University of New York Press.

1990. Conceptual selection. *Philosophical Studies*: 77–87.

1992. An evolutionary account of science: A response to Rosenberg's critical notice. *Biology and Philosophy* 7(2): 229–236.

2003. Darwin's science and Victorian philosophy of science. In *The Cambridge companion to Darwin,* edited by J. Hodge and G. Radick. Cambridge: Cambridge University Press: 168–191.

Hull, David L., and Michael Ruse, eds.

1998. *The philosophy of biology.* Oxford; New York: Oxford University Press.

Hunter Dupree, A.

1968. *Asa Gray 1810–1888.* College ed. Vol. 132. New York: Atheneum. Original edition, Belknap, 1959.

Husserl, Edmund.

1931. *Ideas: General introduction to pure phenomenology (Ideen au einer reinen Phänomenologie und phänomenologischen Philosophie).* Translated by W. R. B. Gibson. New York: Collier Macmillan. Original edition, 1913.

Hutchinson, G. E.

1968. When are species necessary? In *Population biology and evolution,* edited by R. C. Lewontin. Syracuse, NY: Syracuse University Press: 177–186.

Huxley, Julian, ed.

1940. *The new systematics.* London: Oxford University Press.

1942. *Evolution: The modern synthesis.* London: Allen and Unwin.

Huxley, Thomas Henry.

1893a. *Darwiniana: Essays.* London Macmillan.

1893b. Lectures on evolution. In *Collected Essays.* London: Macmillan. Original edition, 1876.

1895. *Man's place in nature and other anthropological essays: Collected Essays by T. H. Huxley.* Vol. 7. London: Macmillan.

1906. *Man's place in nature and other essays.* Everyman's Library ed. London: Dent; New York: Dutton.

Isidore of Seville, Saint.

2005. *Etymologiae: The etymologies of Isidore of Seville.* Translated by S. A. Barney. New York: Cambridge University Press.

Jevons, William Stanley.

1878. *The principles of science: A treatise on logic and scientific method.* 2nd ed. London: Macmillan. Original edition, 1873.

Jordan, D. S.

1905a. The origin of species through isolation. *Science* 22: 545–562.

Jordan, Karl.
    1896. On mechanical selection and other problems. *Novitates Zoologicae* 3: 426–525.
    1905b. Der Gegensatz zwischen geographischer und nichtgeographischer Variation. *Zeitschrift für Wissenschaftliche Zoologie* 83: 151–210.

Jordanova, L. J.
    1984. *Lamarck: Past masters.* Oxford: Oxford University Press.

Joseph, H. W. B.
    1916. *An introduction to logic.* 2nd ed. Oxford: Clarendon Press.

Judd, Walter S., Christopher S. Campbell, Elizabeth A. Kellog, and Peter F. Stevens.
    1999. *Plant systematics: A phylogenetic approach.* Sunderland, MA: Sinauer.

Junker, T.
    1991. Heinrich Georg Bronn and Origin of Species. *Sudhoffs Arch Z Wissenschaftsgesch* 75(2): 180–208.

Jussieu, Antoine-Laurent de.
    1964. *Genera plantarum.* Facsimile ed. Weinheim, Germany: Cramer. Original edition, 1789, Paris.

Kant, Immanuel.
    1933. *Critique of pure reason.* Translated by N. K. Smith. 2nd rev. ed. London: Macmillan. Original edition, 1787, 2nd ed.
    1951. *Critique of judgment.* Translated by J. H. Bernard. New York: Hafner. Original edition, 1790–1793.

Keil, Frank C.
    1995. The growth of causal understandings of natural kinds. In *Causal cognition: A multidisciplinary debate*, edited by D. Sperber, D. Premack, and A. J. Premack. Oxford: Clarendon Press.

Keller, Roberto A., Richard N. Boyd, and Quentin D. Wheeler.
    2003. The illogical basis of phylogenetic nomenclature. *Botanical Review* 69(1): 93–110.

Kitcher, Philip.
    1984. Species. *Philosophy of Science* 51: 308–333.
    1989. Some puzzles about species. In *What the philosophy of biology is: Essays dedicated to David Hull*, edited by M. Ruse. Dordrecht: Kluwer: 183–208.

Kitts, David B.
    1987. Plato on kinds of animals. *Biology and Philosophy* 2(3): 315.

Klima, Guyla.
    2005. The essentialist nominalism of John Buridan. *Review of Metaphysics* 58: 301–315.

Knowles, L. Lacey, and Bryan C. Carstens.
    2007. Delimiting species without monophyletic gene trees. *Systematic Biology* 56(6): 887–895.

Koerner, Lisbet.
  1999. *Linnaeus: Nature and nation.* Cambridge, MA: Harvard University Press.
Kojima, J.
  2003. Apomorphy-based definition also pinpoints a node, and PhyloCode names prevent effective communication. *Botanical Review* 69(1): 44–58.
Kondrashov, A. S.
  1994. The asexual ploidy cycle and the origin of sex. *Nature* 370(6486): 213–216.
Kornet, D.
  1993a. Internodal species concept. *Journal of Theoretical Biology* 104: 407–435.
  1993b. Permanent splits as speciation events: A formal reconstruction of the internodal species concept. *Journal of Theoretical Biology* 164: 407–435.
Kottler, Malcolm J.
  1978. Charles Darwin's biological species concept and theory of geographic speciation: The Transmutation Notebooks. *Annals of Science* 35: 275–297.
Kretzmann, Norman, and Eleonore Stump.
  1988. *Logic and the philosophy of language.* Vol. 1, Cambridge Translations of Medieval Philosophical Texts. Cambridge: Cambridge University Press.
Krüger, Lorenz, Lorraine J. Daston, Michael Heidelberger, Gerd Gigerenzer, and Mary S. Morgan.
  1990. *The Probabilistic revolution.* 2 vols. Cambridge, MA: MIT Press.
Kuhn, Thomas S.
  1959. *The Copernican revolution: Planetary astronomy in the development of Western thought.* New York: Vintage Books.
Kuntz, Marion Leathers, and Paul Grimley Kuntz, eds.
  1988. *Jacob's ladder and the tree of life: Concepts of hierarchy and the Great Chain of Being.* Rev. ed. Vol. 14, American University Studies. Series V, Philosophy. New York: Lang.
Kvist, L., J. Martens, H. Higuchi, A. A. Nazarenko, O. P. Valchuk, and M. Orell.
  2003. Evolution and genetic structure of the great tit *(Parus major)* complex. *Proceedings of the Royal Society of London B, Biological Sciences* 270(1523): 1447–1454.
Lam, H. J.
  1957. What is a taxon? *Taxon* 6 (8): 213–215.
Lamarck, Jean Baptiste.
  1802. *Recherches sur l'organisation des corps vivants.* Paris: Dentu.
  1809. *Philosophie zoologique, ou, Exposition des considérations relative à l'histoire naturelle des animaux.* Paris: Dentu.
  1914. *Zoological philosophy: An exposition with regard to the natural history of animals.* Translated by H. Elliot. London: Macmillan.
Lambert, David M., and Hamish G. Spencer, eds.
  1995. *Speciation and the recognition concept: Theory and application.* Baltimore, MD: Johns Hopkins University Press.
Larson, James L.
  1967. Linnaeus and the Natural Method. *Isis* 58(3): 304–320.

Lawrence, G. H. M., ed.
1963. *Adanson: The bicentennial of Michel Adanson's "Familles des plantes."* Hunt Biological Library. Pittsburgh, PA: Carnegie Institute of Technology.

Lee, J. Y., K. Mummenhoff, and J. L. Bowman.
2002. Allopolyploidization and evolution of species with reduced floral structures in *Lepidium* L. (Brassicaceae). *Proceedings of the National Academy of Sciences of the USA* 99(26): 16835–40.

Lee, James.
1810. *An introduction to the science of botany: Chiefly extracted from the works of Linnaeus; to which are added, several new tables and notes, and a life of the author/by . . . James Lee.* 4th ed. London: Printed for F. C. and J. Rivington; Wilkie and Robinson; J. Walker.

Leff, Gordon.
1958. *Medieval thought from Saint Augustine to Ockham.* Harmondsworth, UK: Penguin.

Leibniz, Gottfried Wilhelm.
1996. *New essays on human understanding.* Translated by P. Remnant and J. Bennett. Cambridge: Cambridge University Press. Original edition, 1765.

Lennox, James G.
1981. Teleology, chance, and Aristotle's theory of spontaneous generation. *Journal of Historical Philosophy* 19: 219–38.
1987. Kinds, forms of kinds, and the more and the less in Aristotle's biology. In *Philosophical issues in Aristotle's biology*, edited by A. Gotthelf and J. G. Lennox. Cambridge: Cambridge University Press: 339–359.
1993. Darwin was a teleologist. *Biology and Philosophy* 8(4): 409–421.
1994. Aristotle's biology: plain, but not simple. *Studies in Historical and Philosophical Science* 25(5): 817–823.
2001. *Aristotle's philosophy of biology: Studies in the origins of life science.* Cambridge: Cambridge University Press.

Lenoir, Timothy.
1980. Kant, Blumenbach, and vital materialism in German biology. *Isis* 71(1): 77–108.
1987. The eternal laws of form: Morphotypes and the conditions of existence in Goethe's biological thought. In *Goethe and the sciences: A re-appraisal*, edited by F. Amrine, F. J. Zucker, and H. Wheeler. Berlin: Springer: 17–28.

Levit, Georgy S., and Kay Meister.
2006. The history of essentialism vs. Ernst Mayr's "Essentialism Story": A case study of German idealistic morphology. *Theory in Biosciences* 124: 281–307.

Lherminer, Philippe, and Michel Solignac.
2000. L'espèce: Définitions d'auters. *Sciences de la Vie* 153–165.

Liddell, H. G., and Scott.
1888. *An intermediate Greek-English lexicon, founded upon the seventh edition of Liddell and Scott's Greek-English Lexicon.* Oxford: Clarendon Press.

Liebers, Dorit, Peter de Knijff, and Andreas J. Helbig.

2004. The herring gull complex is not a ring species *Proceedings of the Royal Society of London B, Biological Sciences* 271: 893–901.

Lindroth, Sten.

1983. The two faces of Linnaeus. In *Linnaeus, the man and his work*, edited by T. Frängsmyr. Berkeley: University of California Press: 1–62.

Linne, Carl von.

1788–1793. *Systema naturae per regna tria naturae, secundum classes, ordines, genera, species, cum characteribus, differentiis, synonymis, locis . . ./cura Jo. Frid. Gmelin.* Editio decima tertia, aucta, reformata. ed. Lipsiae: Georg. Emanuel Beer.

1956. *Caroli Linnaei Systema naturae: A photographic facsimile of the first volume of the tenth edition (1758): Regnum animale.* London: Printed by order of the Trustees, British Museum (Natural History).

Littlejohn, Murray J.

1981. Reproductive isolation: A critical review. In *Evolution and speciation: Essays in honor of M. J. D. White*, edited by W. R. Atchley and D. S. Woodruff. Cambridge: Cambridge University Press: 298–334.

Lotsy, J. P.

1916. *Evolution by means of hybridization.* The Hague: Martinus Nijhoff.

1931. On the species of the taxonomist in its relation to evolution. *Genetica* 13: 1–16.

Lovejoy, Arthur O.

1936. *The great chain of being: A study of the history of an idea.* Cambridge, MA: Harvard University Press.

1959. Buffon and the problem of species. In *Forerunners of Darwin 1745–1859*, edited by B. Glass, O. Temkin, and W. L. Straus. Baltimore, MD: Johns Hopkins University Press: 84–113.

Lucretius.

1969. *On the nature of things (De rerum natura).* Translated by M. F. Smith. London: Sphere Books.

Lurie, Edward.

1960. *Louis Agassiz: A life in science.* Baltimore, MD: Johns Hopkins University Press.

Lyell, Charles.

1832. *Principles of geology, being an attempt to explain the former changes of the earth's surface, by reference to causes now in operation.* 2nd ed. London: Murray.

Mallet, James.

1995. The species definition for the modern synthesis. *Trends in Ecology & Evolution* 10(7): 294–299.

2000. Species and their names. *Trends in Ecology & Evolution* 15(8): 344–345.

2001. Species, concepts of. In *Encyclopedia of biodiversity*, edited by S. A. Levin. New York: Academic Press: 427–440.

2008. Mayr's view of Darwin: Was Darwin wrong about speciation? *Biological Journal of the Linnean Society* 95: 3–16.

Mandelbaum, Maurice.
1957. The scientific background of evolutionary theory in biology. *Journal of the History of Ideas* 18(3): 342–361.

Margulis, Lynn, and Karlene V. Schwartz.
1998. *Five kingdoms: An illustrated guide to the phyla of life on earth.* 3rd ed. San Francisco: Freeman.

Matthews, Gwynneth.
1972. Plato's epistemology and related logical problems. In *Selections from philosophers*, edited by M. Warnock. London: Faber & Faber.

Mayden, R. L.
1997. A hierarchy of species concepts: The denouement in the saga of the species problem. In *Species: The units of diversity*, edited by M. F. Claridge, H. A. Dawah, and M. R. Wilson. London: Chapman and Hall: 381–423.
2002. On biological species, species concepts and individuation in the natural world. *Fish and Fisheries* 3(3): 171–196.

Mayr, Ernst.
1940. Speciation phenomena in birds. *American Naturalist* 74: 249–278.
1942. *Systematics and the origin of species from the viewpoint of a zoologist.* New York: Columbia University Press.
1946. The naturalist in Leidy's time and today. *Proceedings of the Academy of Natural Sciences of Philadelphia* 98: 271–276.
1954. Change of genetic environment and evolution. In *Evolution as a process*, edited by J. Huxley, A. Hardy, and E. Ford. London: Allen and Unwin: 157–180.
1955. Karl Jordan's contribution to current concepts in systematics and evolution. In *Evolution and the diversity of life*, edited by E. Mayr. Cambridge, MA: Harvard University Press: 297–306.
1957. Species concepts and definitions. In *The species problem: A symposium presented at the Atlanta meeting of the American Association for the Advancement of Science, December 28–29, 1955*, Publication No. 50, edited by E. Mayr. Washington, DC: American Association for the Advancement of Science: 1–22.
1959. Darwin and the evolutionary theory in biology. In *Evolution and anthropology: a centennial appraisal*, edited by B. J. Meggers. Washington, DC: Anthropological Society of Washington: 1–10.
1963. *Animal species and evolution.* Cambridge, MA: Belknap Press of Harvard University Press.
1969. *Principles of systematic zoology.* New York: McGraw-Hill.
1970. *Populations, species, and evolution: An abridgment of Animal species and evolution.* Cambridge, MA: Belknap Press of Harvard University Press.
1976. Is the species a class or an individual? *Systematic Zoology* 25: 192.
1980. How I became a Darwinian. In *The evolutionary synthesis*, edited by E. Mayr and W. B. Provine. Cambridge, MA: Harvard University Press: 413–429.

1982. *The growth of biological thought: diversity, evolution, and inheritance.* Cambridge, MA: Belknap Press of Harvard University Press.

1985. The species as category, taxon and population. In *Histoire du concept d'espece dans les sciences de la vie.* Paris: Fondation Singer-Polignac: 303–320.

1988. The why and how of species. *Biology and Philosophy* 3: 431–441.

1991. *One long argument: Charles Darwin and the genesis of modern evolutionary thought.* Cambridge, MA: Harvard University Press.

1992. Species concepts and their application. In *The units of evolution: Essays on the nature of species,* edited by M. Ereshevsky. Cambridge, MA: MIT Press: 15–26.

1994. Ordering systems. *Science* 715–716: 5186.

1995. Species, classification, and evolution. In *Biodiversity and evolution,* edited by R. Arai, M. Kato, and Y. Doi. Tokyo: National Science Museum Foundation: 3–12.

1996. What is a species, and what is not? *Philosophy of Science* 2: 262–277.

1997. *This is biology: The science of the living world.* Cambridge, MA: Belknap Press of Harvard University Press.

1999. *Systematics and the origin of species from the viewpoint of a zoologist.* New York: Columbia University Press. Original edition, 1942.

2000a. The biological species concept. In *Species concepts and phylogenetic theory: A debate,* edited by Q. D. Wheeler and R. Meier. New York: Columbia University Press: 17–29.

2000b. A critique from the biological species concept: What is a species, and what is not? In *Species concepts and phylogenetic theory: A debate,* edited by Q. D. Wheeler and R. Meier. New York: Columbia University Press: 93–100.

Mayr, Ernst, and Peter D. Ashlock.
1991. *Principles of systematic zoology.* 2nd ed. New York: McGraw-Hill.

Mayr, Ernst, E. Gorton Linsley, and Robert L. Usinger.
1953. *Methods and principles of systematic zoology.* New York: McGraw-Hill.

Mayr, Ernst, and William B. Provine.
1980. *The evolutionary synthesis: Perspectives on the unification of biology.* Cambridge, MA: Harvard University Press.

Mazeroll, Anthony I., and Marc Weiss.
1995. The state of confusion in Discus taxonomy. In *Cichlids Yearbook.* N.p.: Cichlid Press: 77–83.

McCulloch, Florence.
1962. *Mediaeval Latin and French bestiaries.* Rev. ed. University of North Carolina Studies in the Romance Languages and Literatures, no. 33. Chapel Hill: University of North Carolina Press.

McKeon, Richard.
1929. *Selections from medieval philosophers.* Vol. 1. New York: Scribner's.
1930. *Selections from medieval philosophers.* Vol. 2. New York: Scribner's.
ed. 1941. *The basic works of Aristotle.* New York: Random House.

McOuat, Gordon R.
1996. Species, rules and meaning: The politics of language and the ends of definitions in 19th century natural history. *Studies in History and Philosophy of Science Part A* 27: 473–519.
2001. Cataloguing power: Delineating "competent naturalists" and the meaning of species in the British Museum. *British Journal for the History of Science* 34: 1–28.
2003. The logical systematist: George Bentham and his *Outline of a New System of Logic. Archives of Natural History* 30(2): 203–223.

Meier, Rudolf, and Rainer Willmann.
2000. The Hennigian species concept. In *Species concepts and phylogenetic theory: A debate*, edited by Q. Wheeler and R. Meier. New York: Columbia University Press.

Mill, John Stuart.
1930. *A system of logic, ratiocinative and inductive: Being a connected view of the principles of evidence and the methods of scientific investigation.* 8th (1860) ed. London: Longmans Green. Original edition, 1843.

Millikan, Ruth Garrett.
1984. *Language, thought, and other biological categories: New foundations for realism.* Cambridge, MA: MIT Press.

Mishler, Brent D., and Robert N. Brandon.
1987. Individuality, pluralism, and the Phylogenetic Species Concept. *Biology and Philosophy* 2: 397–414.

Mishler, Brent D., and Michael J. Donoghue.
1982. Species concepts: A case for pluralism. *Systematic Zoology* 31: 491–503.

Mishler, Brent D., and Edward C. Theriot.
2000. The phylogenetic species concept (*sensu* Mishler and Theriot): Monophyly, apomorphy, and phylogenetic species concepts. In *Species concepts and phylogenetic theory: A debate*, edited by Q. D. Wheeler and R. Meier. New York: Columbia University Press: 44–54.

Morgan, Thomas Hunt.
1903. *Evolution and adaptation.* New York: Macmillan.

Morris, P. J.
1997. Louis Agassiz's additions to the French translation of his *Essay on Classification. Journal of the History of Biology* 30: 121–134.

Morton, A. G.
1981. *History of botanical science: An account of the development of botany from ancient times to the present day.* London: Academic Press.

Moss, Lenny.
2003. *What genes can't do.* Edited by G. McGee and A. Kaplan, *Basic bioethics.* Cambridge MA: Bradford Books, MIT Press.

Muller, H. J.
1940. Bearings of the "*Drosophila*" work on systematics. In *The new systematics*, edited by J. Huxley. London: Oxford University Press: 185–268.

Müller-Wille, Staffan.
2003. Nature as a marketplace: The political economy of Linnaean botany. *History of Political Economy* 35 (annual supplement): 154–172.

Müller-Wille, Staffan, and Vitezslav Orel.
2007. From Linnaean species to Mendelian factors: Elements of hybridism, 1751–1870. *Annals of Science* 64(2): 171–215.

Murray, Desmond.
1955. *Species revalued: A biological study of species as a unit in the economy of nature, shown from plant and insect life.* London: Blackfriars.

Naisbit, R. E., C. D. Jiggins, M. Linares, C. Salazar, and J. Mallet.
2002. Hybrid sterility, Haldane's rule and speciation in *Heliconius cydno* and *H. melpomene*. *Genetics* 161(4): 1517–26.

Nelson, Gareth J.
1971. Paraphyly and polyphyly: Redefinitions. *Systematic Zoology* 20(4): 471–472.
1989. Species and taxa: Speciation and evolution. In *Speciation and its consequences*, edited by D. Otte and J. Endler. Sunderland, MA: Sinauer.

Nelson, Gareth J., and Norman I. Platnick.
1981. *Systematics and biogeography: Cladistics and vicariance.* New York: Columbia University Press.

Nicolson, Adam.
2003. *God's secretaries: The making of the King James Bible.* London: HarperCollins.

Nixon, K. C.
2003. The PhyloCode is fatally flawed, and the "Linnaean" system can easily be fixed. *Botanical Review* 69(1): 111–120.

Nordenskiöld, Erik.
1929. *The history of biology: A survey.* Translated by L. Eyre. London: Kegan Paul, Trench, Trubner.

Numbers, Ronald L.
1992. *The creationists.* New York: Knopf.

Nyhart, Lynn K.
1995. *Biology takes form: Animal morphology and the German universities, 1800–1900.* Chicago: University of Chicago Press.

Oldroyd, David R.
1983. *Darwinian impacts: an introduction to the Darwinian revolution.* 2nd rev. ed. Kensington: University of New South Wales Press.
1986. *The Arch of Knowledge: An introductory study of the history of the philosophy and methodology of science.* Kensington: New South Wales University Press.

Ong, Walter J.
1958. *Ramus, method, and the decay of dialogue: From the art of discourse to the art of reason.* Cambridge, MA: Harvard University Press.

Orr, H. Allen, and Daven C. Presgraves.
2000. Speciation by postzygotic isolation: Forces, genes and molecules. *BioEssays* 22: 1085–1094.

Osborn, Henry Fairfield.
1894. *From the Greeks to Darwin: An outline of the development of the evolution idea.* Columbia University Biological Series. I. New York: Macmillan.

Osborne, Richard H.
1971. *The biological and social meaning of race.* San Francisco: Freeman.

Owen, Richard.
1834. On the Generation of the marsupial animals, with a description of the impregnated uterus of the kangaroo. *Philosophical Transactions of the Royal Society of London* 124: 333–364.

1835. On the osteology of the chimpanzee and orang utan. *Transactions of the Zoological Society* 1: 343–379.

1843. *Lectures on the comparative anatomy and physiology of the invertebrate animals, delivered at the Royal College of Surgeons, in 1843.* London: Longman, Brown, Green, and Longmans.

1848. *The archetype and homologies of the vertebrate skeleton.* London: van Voorst.

1855. *Lectures on the comparative anatomy and physiology of the invertebrate animals: Delivered at the Royal College of Surgeons.* 2nd ed. London: Longman, Brown, Green and Longmans.

Padian, Kevin.
1999. Charles Darwin's view of classification in theory and practice. *Systematic Biology* 48(2): 352–364.

Panchen, Alec L.
1992. *Classification, evolution, and the nature of biology.* Cambridge: Cambridge University Press.

Paris, C. A., F. S. Wagner, and W. H. Wagner Jr.
1989. Cryptic species, species delimitation, and taxonomic practice in the homosporous ferns. *American Fern Journal* 79(2): 46–54.

Paterson, Hugh E. H.
1985. The recognition concept of species. In *Species and speciation,* edited by E. Vrba. Pretoria: Transvaal Museum: 21–29.

1993. *Evolution and the recognition concept of species. Collected writings.* Baltimore, MA: John Hopkins University Press.

Paul, Diane.
1981. "In the interests of civilization": Marxist views of race and culture in the nineteenth century. *Journal of the History of Ideas* 42(1): 115–138.

Pavord, Anna.
2005. *The naming of names: The search for order in the world of plants.* New York: Bloomsbury.

Peirce, Charles Sanders.
　1885. On the algebra of logic: A contribution to the philosophy of notation. *American Journal of Mathematics* 7: 180–202.

Pellegrin, Pierre.
　1982. *La classification des animaux chez Aristote: Statut de la biologie et unité de l'aristotelisme*. Paris: Société d'edition "Les Belles lettres."
　1986. *Aristotle's classification of animals: Biology and the conceptual unity of the Aristotelian corpus*. Translated by A. Preus. Rev. ed. Berkeley: University of California Press.
　1987. Logical difference and biological difference: The unity of Aristotle's thought. In *Philosophical issues in Aristotle's biology*, edited by A. Gotthelf and J. G. Lennox. Cambridge: Cambridge University Press: 313–338.

Pennisi, E.
　2002. Ecology. A coral by any other name. *Science* 296(5575): 1949–50.

Plato.
　1998. *Phaedrus*. Translated by J. H. Nichols. Ithaca, NY: Cornell University Press.

Pleijel, Frederik.
　1999. Phylogenetic taxonomy, a farewell to species, and a revision of *Heteropodarke (Hesionidae, Polychaeta, Annelida)*. *Systematic Biology* 48(4): 755–789.

Pleijel, Frederik, and G. W. Rouse.
　2000. Least-inclusive taxonomic unit: A new taxonomic concept for biology. *Proceedings of the Royal Society of London B, Biological Sciences* 267(1443): 627–630.
　2003. Ceci n'est pas une pipe: Names, clades and phylogenetic nomenclature. *Journal of Zoological Systematics and Evolutionary Research* 41(3): 162–174.

Pliny the Elder.
　1906. *Naturalis historia*. Edited by K. F. T. Mayhoff. Lipsiae: Teubner.
　1940–1963. *Natural history*. Translated by H. Rackham, D. E. Eichholz, and W. H. S. Jones. Vol. 1–10. London: Heinemann.

Polly, Paul David.
　1997. Ancestry and species definition in paleontology: A stratocladistic analysis of Paleocene-Eocene Viverravidae (Mammalia, Carnivora) from Wyoming. *Contributions from the Museum of Paleontology, The University of Michigan* 30(1): 1–53.

Popper, Karl R.
　1957a. *The open society and its enemies*. 3rd ed. London: Routledge and Kegan Paul.
　1957b. *The poverty of historicism*. London: Routledge and Kegan Paul.
　1959. *The logic of scientific discovery*. Translated by K. Popper, J. Freed, and L. Freed. London: Hutchinson.
　1960. *The poverty of historicism*. 2nd ed. London: Routledge and Kegan Paul.

Porphyry, and Jonathan Barnes.
2003. *Porphyry's introduction.* Clarendon Later Ancient Philosophers. Oxford: Oxford University Press.

Porphyry, the Phoenician.
1975. *Isagoge.* Translated by E. W. Warren. Toronto: Pontifical Institute of Mediaeval Studies.

Poulton, Edward Bagnall.
1903. What is a species? *Proceedings of the Entomological Society of London.* Reprinted in Poulton 1908: 46–94.
1908. *Essays on evolution.* Oxford: Clarendon.

Preus, A.
2002. Plotinus and biology. In *Neoplatonism and nature: Studies in Plotinus' Enneads,* edited by M. F. Wagner. Albany: State University of New York Press: 43–55.

Prichard, James Cowles.
1813. *Researches into the physical history of man.* London: Houlston & Stoneman.

Quine, Willard Van Ormand.
1960. *Word and object: Studies in communication.* Cambridge: Technology Press of the Massachusetts Institute of Technology.

Ramsbottom, John.
1938. Linnaeus and the species concept. *Proceedings of the Linnean Society of London* 150: 192–220.

Ramsey, J., D. W. Schemske, and J. A. Doyle.
1998. Pathways, mechanisms, and rates of polyploid formation in flowering plants: Phylogeny of vascular plants. *Annual Review of Ecology and Systematics* 29: 467–501, 567–599.

Raven, Charles Earle.
1953. *Natural religion and Christian theology: Science and Religion.* Vol. I. Cambridge: Cambridge University Press.
1986. *John Ray, naturalist: His life and works.* 2nd ed. Cambridge: Cambridge University Press.

Regan, C. Tate.
1926. Organic evolution. *Report of the British Association for the Advancement of Science* 1925: 75–86.

Rensch, Bernhard.
1928. Grenzfälle von Rasse und Art. *Journal für Ornithologie* 76: 222–231.
1929. *Das Prinzip geographischer Rassenkreise und das Problem der Artbildung.* Berlin: Gebrüder Borntraeger.
1947. *Neuere Probleme der Abstammungslehre.* Stuttgart: Enke.
1959. *Evolution above the species level.* New York: Columbia University Press. Original edition, 2nd ed., of *Neuere Probleme der Abstammungslehre,* Ferdinand Enke, Stuttgart, 1954.
1980. Historical development of the present Synthetic Neo-Darwinism in Germany. In *The evolutionary synthesis,* edited by E. Mayr and W. B. Provine. Cambridge, MA: Harvard University Press: 284–303.

Richards, Robert J.
  2000. Kant and Blumenbach on the Bildungstrieb: A historical misunderstanding. *Studies in Historical, Philosophical, Biological & Biomedical Sciences* 31(1): 11–32.

Rieppel, Olivier.
  2003. Semaphoronts, cladograms and the roots of total evidence. *Biological Journal of the Linnean Society* 80(1): 167–186.

Rieseberg, Loren H., and J. M. Burke.
  2001. A genic view of species integration. *Journal of Evolutionary Biology* 14(6): 883–886.

Robson, Guy Coburn.
  1928. *The species problem: An introduction to the study of evolutionary divergence in natural populations.* Biological Monographs and Manuals, no. 8. Edinburgh: Oliver and Boyd.

Roger, Jacques.
  1997. *Buffon: A life in natural history.* Translated by S. L. Bonnefoi. Edited by L. P. Williams. Cornell History of Science Series. Ithaca, NY: Cornell University Press.

Romanes, George John.
  1895. *Darwin, and after Darwin: An exposition of the Darwinian theory and a discussion of the post-Darwinian questions.* 3 vols. Vol. II: *Post-Darwinian questions: Heredity and utility.* London: Longmans, Green.

Rosen, Donn E.
  1978. Vicariant patterns and historical explanation in biogeography. *Systematic Zoology* 27: 159–188.
  1979. Fishes from the uplands and intermontane basins of Guatemala: Revisionary studies and comparative biogeography. *Bulletin of the American Museum of Natural History* 162: 267–376.

Ross, David.
  1949. *Aristotle.* 5th ed. London: Methuen/University Paperbacks.

Rossi, Paolo.
  2000. *Logic and the art of memory: The quest for a universal language.* London: Athlone.

Runes, Dagobert D, ed.
  1962. *Classics in logic: Readings in epistemology, theory of knowledge and dialectics.* New York: Philosophical Library.

Ruse, M.
  1987. Biological species: Natural kinds, individuals, or what? *British Journal for the Philosophy of Science* 38: 225–242.
  1998. All my love is toward individuals. *Evolution* 52: 283–288.

Russell, E. S.
  1982. *Form and function: A contribution to the history of animal morphology.* Chicago: University of Chicago Press. Original edition, 1916, Murray.

Sachs, J. V.
  1890. *History of botany (1530–1860)*. Translated by H. E. F. Garnsey and I. B. Balfour. Oxford: Clarendon Press. Original edition, 1875.

Salisbury, E. J.
  1940. Ecological aspects of plant taxonomy. In *The new systematics*, edited by J. Huxley. London: Oxford University Press.

Salthe, Stanley N.
  1985. *Evolving hierarchical systems: Their structure and representation.* New York: Columbia University Press.

Salzburger, W., S. Baric, and C. Sturmbauer.
  2002. Speciation via introgressive hybridization in East African cichlids? *Molecular Ecology* 11(3): 619–25.

Sankey, Howard.
  1998. Taxonomic incommensurability. *International Studies in the Philosophy of Science* 12(1): 7–16.

Santayana, George.
  1917. *The life of reason, or the phases of human progress.* Vol. 1. New York: Scribner's. Original edition, 1905.

Schilthuizen, Menno.
  2000. Dualism and conflicts in understanding speciation. *BioEssays* 22: 1134–1141.

Schloegel, Judy Johns.
  1999. From anomaly to unification: Tracy Sonneborn and the species problem in protozoa, 1954–1957. *Journal of the History of Biology* 32(1): 93–132.

Schopf, Thomas J. M.
  1972. *Models in paleobiology.* San Francisco: Freeman Cooper.

Scoble, M. J.
  1985. The species in systematics. In *Species and speciation*, edited by E. Vrba. Pretoria: Transvaal Museum: 31–34.

Scott, Walter, ed.
  1924. *Hermetica: The ancient Greek and Latin writings which contain religious or philosophic teachings ascribed to Hermes Trimegistus.* Boulder, CO: Hermes House.

Sedley, D. N.
  2007. *Creationism and its critics in antiquity.* Sather Classical Lectures. Berkeley: University of California Press.

Senn, G. 1925.
  Die Einführung des Art- und Gattungsbegriffs in die Biologie. *Verhandlung der Schweizer. Naturforsh. Gesellschaftung* II: 183–184.

Simpson, George Gaylord.
  1943. Criteria for genera, species and subspecies in zoology and paleontology. *Annals New York Academy of Science* 44: 145–178.
  1944. *Tempo and mode in evolution.* New York: Columbia University Press.

1951. The species concept. *Evolution* 5: 285–298.

1961. *Principles of animal taxonomy*. New York: Columbia University Press.

Simpson, James Y.

1925. *Landmarks in the struggle between science and religion*. London: Hodder and Stoughton.

Singer, Charles Joseph.

1950. *A history of biology to about the year 1900: A general introduction to the study of living things*. 2nd ed. London: Abelard-Schuman.

Slaughter, Mary M.

1982. *Universal languages and scientific taxonomy in the seventeenth century*. Cambridge: Cambridge University Press.

Sloan, Phillip R.

1979. Buffon, German biology, and the historical interpretation of biological species. *British Journal for the History of Science* 12(41): 109–153.

1985. From logical universals to historical individuals: Buffon's idea of biological species. In *Histoire du concept d'espece dans les sciences de la vie*. Paris: Fondation Singer-Polignac: 101–140.

Smith, Andrew B.

1994. *Systematics and the fossil record: Documenting evolutionary patterns*. Oxford: Blackwell Science.

Smith, James Edward.

1821. *A grammar of botany, illustrative of artificial, as well as natural, classification, with an explanation of Jussieu's system*. London: Longman, Hurst, Rees, Orme, and Brown.

Sneath, P. H. A., and Robert R. Sokal.

1973. *Numerical taxonomy: The principles and practice of numerical classification*. San Francisco: Freeman.

Sober, Elliott.

1980. Evolution, population thinking, and essentialism. *Philosophy of Science* 47: 350–383.

1984. *The nature of selection: Evolutionary theory in philosophical focus*. Cambridge, MA: MIT Press.

Sokal, Robert R., and T. Crovello.

1970. The biological species concept: A critical evaluation. *American Naturalist* 104: 127–153.

Sokal, Robert R., and P. H. A. Sneath.

1963. *Principles of numerical taxonomy*. San Francisco: Freeman.

Sonneborn, T. M.

1957. Breeding systems, reproductive methods and species problems in Protozoa. In *The species problem: A symposium presented at the Atlanta meeting of the American Association for the Advancement of Science, December 28–29, 1955*, Publication No. 50, edited by E. Mayr. Washington, DC: American Association for the Advancement of Science: 155–324.

Soong, K., and J. C. Lang.

1992. Reproductive integration in reef corals. *Biological Bulletin* 183(3): 418–431.

Sperber, Dan, David Premack, and Ann James Premack, eds.

1995. *Causal cognition: A multidisciplinary debate.* Oxford: Clarendon Press.

Sperber, Dan, and Deirdre Wilson.

1986. *Relevance: Communication and cognition.* Oxford UK: Blackwell.

Sprague, T. A., and E. Nelmes.

1928–1931. The herbal of Leonhart Fuchs. *Journal of the Linnean Society, Botany* 48: 545–642.

Spruit, Leen.

1994–1995. *Species intelligibilis: From perception to knowledge.* 2 vols. Leiden: Brill.

Stadler, P. F., and J. C. Nuno.

1994. The influence of mutation on autocatalytic reaction networks. *Mathematical Biosciences* 122(2): 127–60.

Stafleu, Franz Antonie.

1963. Adanson and the "Familles des plantes." In *Adanson: The bicentennial of Michel Adanson's "Familles des plantes,"* edited by G. H. M. Lawrence. Pittsburgh, PA: Carnegie Institute of Technology: 123–264.

1971. *Linnaeus and the Linnaeans: The spreading of their ideas in systematic botany, 1735–1789, Regnum vegetabile, vol. 79.* Utrecht: Oosthoek.

Stamos, David N.

1998. Buffon, Darwin, and the non-individuality of species: A reply to Jean Gayon. *Biology and Philosophy* 13(3): 443–470.

Stannard, Jerry.

1968. Medieval reception of classical plant names. In *Actes du XIIe Congrès International d'Histoire des Sciences, 21–31 Août 1968, Paris.* Paris: Congrès International d'Histoire des Sciences.

1979. Identification of the plants described by Albertus Magnus' *De vegetabilibus* lib. VI. *Res Publica Litterarum* 2: 281–318.

1980. Albertus Magnus and medieval herbalism. In *Albertus Magnus and the sciences: Commemorative essays,* edited by J. A. Weisheipl. Toronto: Pontifical Institute of Mediaeval Studies.

Stannard, Jerry, Richard Kay, and Katherine E. Stannard.

1999. *Herbs and herbalism in the Middle Ages and Renaissance.* Aldershot, UK: Ashgate Variorum.

Stauffer, J. R., K. R. McKaye, and A. F. Konings.

2002. Behaviour: An important diagnostic tool for Lake Malawi cichlids. *Fish and Fisheries* 3(3): 213–224.

Sterelny, Kim.

1999. Species as evolutionary mosaics. In *Species: New interdisciplinary essays,* edited by R. A. Wilson. Cambridge, MA: Bradford/MIT Press: 119–138.

Stevens, Peter F.

1994. *The development of biological systematics: Antoine-Laurent de Jussieu, nature, and the natural system.* New York: Columbia University Press.

1997. J. D. Hooker, George Bentham, Asa Gray and Ferdinand Mueller on species limits in theory and practice: A mid–nineteenth century debate and its repercussions. *Historical Records of Australian Science* 11(3): 345–370.

Strawson, Peter Frederick.

1964. *Individuals: An essay in descriptive metaphysics.* London: Methuen.

Stresemann, Erwin.

1919. Über die euopäischen Baumläufer. *Verhandlungen der Ornithologischen Gesellschaft in Bayern* 14: 39–74.

1975. *Ornithology from Aristotle to the present.* Translated by H. J. Epstein and C. Epstein. Cambridge MA: Harvard University Press.

Strickland, Hugh. E., John Phillips, John Richardson, Richard Owen, Leonard Jenyns, William J. Broderip, John S. Henslow, William E. Shuckard, George R. Waterhouse, William Yarrell, Charles R. Darwin, and John O. Westwood.

1843. Report of a committee appointed "to consider of the rules by which the nomenclature of zoology may be established on a uniform and permanent basis." *Report of the British Association for the Advancement of Science for 1842*: 105–121.

Suppe, Frederick.

1977. The search for philosophic understanding of scientific theories. In *The structure of scientific theories*, edited by F. Suppe. Urbana: University of Illinois Press: 3–232.

1988. *The semantic conception of theories and scientific realism.* Urbana: University of Illinois Press.

Swainson, William.

1834. *Preliminary discourse on the study of natural history.* London: Longman, Rees, Orme, Brown, Green and Longman.

Sylvester-Bradley, Peter Colley, ed.

1956. *The species concept in palaeontology: A symposium*, Publication of the Systematics Association, no. 2. London: Systematics Association.

Syvanen, Michael, and Clarence I. Kado, eds.

1998. *Horizontal gene transfer.* Boston: Kluwer Academic.

Szathmáry, Eörs.

1992. Viral sex, levels of selection, and the origin of life. *Journal of Theoretical Biology* 159(1): 99–109.

Taylor, J. W., D. J. Jacobson, and M. C. Fisher.

1999. The evolution of asexual fungi: Reproduction, speciation and classification. *Annual Review of Phytopathology* 37: 197–246.

Taylor, Peter D.

1992. Community. In *Keywords in evolutionary biology*, edited by E. F. Keller and E. A. Lloyd. Cambridge MA: Harvard University Press: 52–60.

Temkin, Oswei.
   1959. The idea of descent in post-Romantic German biology, 1848–1858. In *Forerunners of Darwin 1745–1859*, edited by B. Glass, O. Temkin, and W. L. Straus Jr. Baltimore, MD: Johns Hopkins University Press: 323–355.

Templeton, Alan R.
   1989. The meaning of species and speciation: A genetic perspective. In *Speciation and its consequences*, edited by D. Otte and J. Endler. Sunderland, MA: Sinauer: 3–27.

Theophrastus.
   1916. *Enquiry into plants, and minor works on odours and weather signs.* Translated by A. Hort. Loeb Classical Library. London: Heinemann.
   1929. *Metaphysics.* Translated by W. D. Ross and F. H. Fobes. Oxford: Clarendon Press.

Thompson, J. Arthur.
   1934. *Biology for everyman.* 2 vols. London: Dent.

Thompson, William R.
   1958. Introduction. In *The Origin of Species, Everyman Edition.* London: J. M. Dent. Original edition, 1928.
   1971. The status of species. *Studia Entomologica* 14 (Novembro): 399–456.

Thorpe, W. H.
   1973. William Robin Thompson. 1887–1972. *Biographical Memoirs of Fellows of the Royal Society* 19: 654–678.

Toulmin, S.
   1972. *Human understanding.* Cambridge: Cambridge University Press.

Tournefort, Joseph Pitton, de.
   1716–30. *The compleat herbal: Or, the botanical institutions of Mr. Tournefort.* Translated by J. Martyn. London: R. Bonwicke. Original edition, *Institutiones rei herbariae*, 1700.

Trémaux, Pierre.
   1865. *Origin et transformations de l'homme et des autres êtres.* Paris: Hachette.
   1874. *Origine des espèces et de l'homme, avec les causes de fixité et de transformation, et principe universel du mouvement et de la vie ou loi des transmissions de force.* 4th 1878 ed. Bale, Switzerland: Librairie Française.

Trewavas, E.
   1973. What Tate Regan said in 1925. *Systematic Zoology* 22: 92–93.

Turesson, Göte.
   1922a. The genotypical response of the plant species to the habitat. *Hereditas* 3: 211–350.
   1922b. The species and variety as ecological units. *Hereditas* 3: 10–113.
   1925. The plant species in relation to habitat and climate. *Hereditas* 6: 147–236.
   1927. Contributions to the genecology upon plant species. *Hereditas* 9: 81–101.
   1929. Zur natur und begrenzung der artenheiten. *Hereditas* 12: 323–334.
   1930. The selective effect of climate upon plant species. *Hereditas* 14: 99, 274, 300.

Turney, Peter, Darrell Whitley, and Russell Anderson.
    1996. Evolution, learning, and instinct: 100 years of the Baldwin Effect. *Evolutionary Computation* 4(3).

Turrill, W. B.
    1940. Experimental and synthetic plant taxonomy. In *The new systematics*, edited by J. Huxley. London: Oxford University Press: 47–72.

Van Alphen, J. J. M., and O. Seehausen.
    2001. Sexual selection, reproductive isolation and the genic view of speciation. *Journal of Evolutionary Biology* 14(6): 874–875.

Van Valen, L.
    1976. Ecological species, multispecies, and oaks. *Taxon* 25: 233–239.

Veron, J. E. N.
    2001. Reticulate evolution in corals. Unpublished manuscript.

Voegelin, Eric.
    1998. *The history of the race idea: From Ray to Carus*. Baton Rouge: Louisiana State University Press.

Vogel, J. C., Steve J. Russell, John A. Barrett, and Mary Gibby.
    1996. A non-coding region of chloroplast DNA as a tool to investigate reticulate evolution in European *Asplenium*. In *Pteridology in perspective*, edited by J. M. Camus, M. Gibby and R. J. Johns. Kew, UK: Royal Botanic Gardens: 313–327.

Vogler, A. P.
    2001. The genic view: A useful model of the process of speciation? *Journal of Evolutionary Biology* 14(6): 876–877.

Vollmer, S. V., and S. R. Palumbi.
    2002. Hybridization and the evolution of reef coral diversity. *Science* 296(5575): 2023–2025.

Vrana, P., and Ward Wheeler.
    1992. Individual organisms as terminal entities: Laying the species problem to rest. *Cladistics* 8: 67–72.
    1911. *The Mutation Theory: Experiments and observations on the origin of species in the vegetable kingdom*. London: Kegan Paul, Trench, Trubner.

Wagner, Moritz.
    1889. *Die Entstehung der Arten durch raumliche Sonderung*. Basel: Benno Schwalbe.

Wagner, Warren H.
    1983. Reticulistics: The recognition of hybrids and their role in cladistics and classification. In *Advances in cladistics*, edited by N. I. Platnick and V. A. Funk. New York: Columbia University Press: 63–79.

Wagner, Warren H., and Alan R. Smith.
    1999. Pteridophytes of North America. In *Flora of North America*. Cambridge, MA: Flora of North America Association.

Wallace, Alfred Russel.
    1858. Note on the theory of permanent and geographical varieties. *The Zoologist* 16: 5887–5888.

1870. *Contributions to the theory of natural selection: A series of essays.* London: Macmillan.

1889. *Darwinism: An exposition of the theory of natural selection, with some of its applications.* London: Macmillan.

Wasmann, Erich.
1910. *Modern Biology and the Theory of Evolution.* Translated by A. M. Buchanan. 3rd ed. London: Kegan Paul, Trench, Trübner. Original edition, 1906.

Weismann, August.
1904. *The evolution theory.* Translated by J. A. Thompson and M. R. Thompson. 2 vols. London: Arnold.

Whately, Richard.
1875. *Elements of logic.* 9th ed. London: Longmans, Green. Original edition, 1826.

Wheeler, Q. D.
1999. Why the phylogenetic species concept? Elementary. *Journal of Nematology* 31(2): 134–141.

Wheeler, Quentin D., and Rudolf Meier, eds.
2000. *Species concepts and phylogenetic theory: A debate.* New York: Columbia University Press.

Wheeler, Quentin D., and Norman I. Platnick.
2000. The phylogenetic species concept (*sensu* Wheeler and Platnick). In *Species concepts and phylogenetic theory: A debate,* edited by Q. D. Wheeler and R. Meier. New York: Columbia University Press: 55–69.

Whewell, William.
1831. Review of Herschel's *Preliminary Discourse* (1830). *Quarterly Review* 45: 374–407.
1837. *History of the inductive sciences.* London: Parker.

White, T. H.
1954. *The book of beasts: Being a translation from a Latin bestiary of the twelfth century.* London: Cape.

Whitehead, Alfred North.
1938. *Science and the modern world.* Harmondsworth, UK: Penguin.

Whitehead, Alfred North, and Bertrand Russell.
1910. *Principia mathematica.* Cambridge: Cambridge University Press.

Wiley, E. O.
1978. The evolutionary species concept reconsidered. *Systematic Zoology* 27: 17–26.
1981. *Phylogenetics: The theory and practice of phylogenetic systematics.* New York: Wiley.

Wiley, E. O., and Richard L. Mayden.
2000. The evolutionary species concept. In *Species concepts and phylogenetic theory: A debate,* edited by Q. D. Wheeler and R. Meier. New York: Columbia University Press: 70–89.

Wilke, C. O., J. L. Wang, C. Ofria, R. E. Lenski, and C. Adami.
  2001. Evolution of digital organisms at high mutation rates leads to survival of the flattest. *Nature* 412(6844): 331–333.

Wilkins, John.
  1970. *The mathematical and philosophical works of the Right Rev. John Wilkins.* 2nd ed. London: Cass. Original edition, 1802.

Wilkins, John S.
  1998. The evolutionary structure of scientific theories. *Biology and Philosophy* 13(4): 479–504.
  1999. On choosing to evolve: Strategies without a strategist. *Journal of Memetics: Evolutionary Models of Information Transmission* 3, http://www.cpm.mmu.ac.uk/jom-emit/1999/vol3/wilkins_j2.html.
  2001. The appearance of Lamarckism in the evolution of culture. In *Darwinism and evolutionary economics*, edited by J. Laurent and J. Nightingale. Cheltenham, UK: Elgar: 160–183.
  2002. Darwinism as metaphor and analogy: Language as a selection process. *Selection: Molecules, Genes, Memes* 3(1): 57–74.
  2003. How to be a chaste species pluralist-realist: The origins of species modes and the Synapomorphic Species Concept. *Biology and Philosophy* 18: 621–638.
  2005. A scientific modern amongst medieval species. *University of Queensland Historical Proceedings* 16: 1–5.
  2007a. The concept and causes of microbial species. *Studies in History and Philosophy of the Life Sciences* 28(3): 389–408.
  2007b. The dimensions, modes and definitions of species and speciation. *Biology and Philosophy* 22(2): 247–266.
  2008. The adaptive landscape of science. *Biology and Philosophy* 659–671.

Wilkins, John S., and Gareth J. Nelson.
  2008. Trémaux on species: A theory of allopatric speciation (and punctuated equilibrium) before Wagner. *PhilSci Archive*, http://philsci-archive.pitt.edu/archive/00003881/, accessed March 9, 2008.

Williams, David M., and T. Martin Embley.
  1996. Microbial diversity: domains and kingdoms. *Annual Review of Ecology and Systematics* 27: 569–595.

Williams, George C.
  1966. *Adaptation and natural selection: A critique of some current evolutionary thought.* Princeton, NJ: Princeton University Press.

Willmann, Rainer.
  1985a. *Die Art in Raum und Zeit.* Berlin: Parey.
  1985b. Reproductive isolation and the limits of the species in time. *Cladistics* 2: 336–338.
  1997. Phylogeny and the consequences of molecular systematics. In *Ephemeroptera and Plecoptera: Biology-ecology-systematics*, edited by P. Landolt and M. Satori. Fribourg: MTL.

Wilson, Leonard G., ed.
    1970. *Sir Charles Lyell's scientific journals on the species question.* Vol. 5, Yale Studies in the History of Science and Medicine. New Haven, CT: Yale University Press.

Windelband, Wilhelm.
    1900. *History of ancient philosophy.* Translated by H. E. Cushman. New York: Scribner's.

Winsor, Mary Pickard.
    1979. Louis Agassiz and the species question. *Studies in History of Biology* 3: 89–117.
    2000. Species, demes, and the Omega Taxonomy: Gilmour and The New Systematics. *Biology and Philosophy* 15(3): 349–388.
    2001. Cain on Linnaeus: the scientist-historian as unanalysed entity. *Studies in the History and Philosophy of the Biological and Biomedical Sciences* 32(2): 239–254.
    2003. Non-essentialist methods in pre-Darwinian taxonomy. *Biology & Philosophy* 18: 387–400.
    2004. Setting up milestones: Sneath on Adanson and Mayr on Darwin. In *Milestones in systematics: Essays from a symposium held within the 3rd Systematics Association Biennial Meeting, September 2001,* edited by D. M. Williams and P. L. Forey. London: Systematics Association: 1–17.

Winston, Judith E.
    1999. *Describing species: Practical taxonomic procedures for biologists.* New York: Columbia University Press.

Wirtjes, Hanneke.
    1991. *The Middle English Physiologus.* Oxford: Oxford University Press.

Woese, C. R.
    1998. Default taxonomy: Ernst Mayr, view of the microbial world. *Proceedings of the National Academy of Sciences of the USA* 95(19): 11043–11046.

Wollheim, Richard.
    1968. *Art and its objects: An introduction to aesthetics.* New York: Harper & Row.

Wood, Casey A., and Florence Marjorie Fyfe, eds.
    1943. *The art of falconry, being the De arte venandi cum avibus of Frederick II of Hohenstaufen.* Stanford, CA: Stanford University Press.

Woodger, J. H.
    1937. *The axiomatic method in biology.* Cambridge: Cambridge University Press.
    1952. From biology to mathematics. *British Journal for the Philosophy of Science* 3: 1–21.

Wright Henderson, P. A.
    1910. *The life and times of John Wilkins.* Edinburgh: Blackwood.

Wright, Sewall.

　1931. Evolution in Mendelian populations. *Genetics* 16(2): 97–159.

　1932. The roles of mutation, inbreeding, crossbreeding and selection in evolution. In *Proceedings of the Sixth International Congress of Genetics*, edited by D. F. Jones. Brooklyn, NY: Brooklyn Botanic Garden: 356–366.

Wu, Chung-I.

　2001. The genic view of the process of speciation. *Journal of Evolutionary Biology* 14: 851–865.

Yatskievych, G., and R. C. Moran.

　1989. Primary divergence and species concepts in ferns. *American Fern Journal* 79(2): 36–45.

Zirkle, Conway.

　1959. Species before Darwin. *Proceedings of the American Philosophical Society* 103(5): 636–644.

Index

Greene, John, 7
Green-Pedersen, Niels Jorgen , 37
Grene, Marjorie, viii
Grew, Nehemiah, 68–69
gross dynastic relationship
    (GDYN), 211
group selectionism, 149

Hacking, Ian, viii
Haeckel, Ernst Heinrich Philipp
    August, 114, 125, 165, 206,
    240n10
Hagberg, Knut, 73, 240n9
Hamilton, William, 65, 101
haplotype groups, 215–216
Harlan, J. R., 220
Hartmann, Nicolai, 236n8
Hennig, Willi, 154, 205, 206
Hennig Convention, 206, 210–211
Hennigian species concept, 206, 207f,
    208–211
Henslow, John Stevens, 134–135
Heraclitus, 12–13
herbals, 38, 57, 58
Hermann, Johannes, 84
Hermetic tradition, 27–28
Hey, Jody, 6, 221, 222
Histoire naturelle (Buffon), 75, 76, 77
Histoire naturelle des oiseaux
    (Buffon), 78
Historia animalium (Gesner), 55
Historia naturalis (Pliny the Elder), 28
Historia plantarum generalis (Ray), 66
Historia plantarum (Gesner), 55
Historia plantarum (Ray and
    Willughby), 65
History of Animals (Aristotle), 18,
    19, 21
"history of ideas," 7–8
history of species concept, 9
holophyly, 248n6
Homo erectus, 225
homology, 54, 116, 135, 156
Homo sapiens, 71
Hooker, Joseph, 125–126, 133
horses, 44
Hort, Arthur, 22–23
Hugh of Fouilloy, 38, 40

Hull, David L., viii, 2, 3–4, 15, 16,
    100–101, 192, 193, 198, 203,
    220, 236n10, 249n1
human race. *See* race
Husserl, Edmund, 102–103
Hutchinson, G. E., 217, 220
Huxley, Julian, 1, 181, 186–187, 222
Huxley, T. H., 165–167
Huxley, Thomas, 94, 135
Hyatt, Alpheus, 171
hybrids and hybridization
    Adanson, 81
    Albertus Magnus, 44
    Aristotle, 21
    Bacon, 59–60
    Buffon, 78
    Candolle, 122, 220, 230
    compilospecies, 220
    Dana, 115–116
    Darwin, 131–132, 150–152,
        157, 230
    Fisher, 183
    fixism, 93, 94, 95
    Gray, 122–123
    Huxley, 187
    Linnaeus, 70, 72, 73, 74
    Lotsy, 174–176
    Lyell, 120–121
    as monsters, 91
    nothospecies, 220
    Quercus species, 217
    Ray, 66
    Romanes, 167
    Wallace, 161
    See also interbreeding
hyenas, 39

ideai, 15
ideal morphologists, 82–85, 114,
    116–118
Ideas (Husserl), 102–103
immutability. *See* mutability and
    immutability
inductive reasoning, 68
infertility
    Dana, 115
    Darwin, 150–152, 157
    Huxley, 166

About the Author

John S. Wilkins is a research fellow in the Department of Philosophy at the University of Sydney. His research focuses on the relation between evolution and religion, the philosophy of taxonomy, and the history of biology. He has also published in the areas of cultural evolution, the philosophy of science, and on science communication through the new technologies. Evolving Thoughts (http://evolvingthoughts.net), his popular blog on science and philosophy of biology, is highly regarded.

Indexer:	Indexing Solutions
Composition:	Michael Bass Associates
Text:	10/13 Sabon
Display:	Franklin Gothic
Printer and Binder:	Thomson-Shore, Inc.